T0146341

Consolidated Afloat Networks and Enterprise Services (CANES)

Manpower, Personnel, and Training Implications

Harry J. Thie, Margaret C. Harrell, Aine Seitz McCarthy, Joseph Jenkins

Prepared for the United States Navy

Approved for public release; distribution unlimited

NATIONAL DEFENSE RESEARCH INSTITUTE

The research described in this report was prepared for the United States Navy. The research was conducted in the RAND National Defense Research Institute, a federally funded research and development center sponsored by the Office of the Secretary of Defense, the Joint Staff, the Unified Combatant Commands, the Department of the Navy, the Marine Corps, the defense agencies, and the defense Intelligence Community under Contract W74V8H-06-C-0002.

Library of Congress Cataloging-in-Publication Data is available for this publication.

978-0-8330-4885-1

Published 2009 by the RAND Corporation
1776 Main Street, P.O. Box 2138, Santa Monica, CA 90407-2138
1200 South Hayes Street, Arlington, VA 22202-5050
4570 Fifth Avenue, Suite 600, Pittsburgh, PA 15213-2665
RAND URL: http://www.rand.org/
To order RAND documents or to obtain additional information, contact
Distribution Services: Telephone: (310) 451-7002;
Fax: (310) 451-6915; Email: order@rand.org

Preface

The average age of a typical Navy shipboard network is about seven years. These networks and the systems and applications that reside on them are an amalgam of disparate hardware and operating software that were developed and introduced onboard largely independent from one another. The Consolidated Afloat Networks and Enterprise Services (CANES) initiative is designed to consolidate and improve the networks on tactical platforms, largely through a common computing environment. The Navy Program Executive Officer for Command, Control, Communications, Computers, and Intelligence (PEO C4I) asked the RAND National Defense Research Institute to assess broadly the manpower, personnel, and training implications associated with the introduction of CANES on naval ships. The research identified the implications of the conversion from current systems to understand the impact on the numbers and types of personnel needed and on overall training demands. This report describes the results of this research. It should be of interest to those involved in the analysis and policy planning for fleet manpower determination, personnel management, and training management.

This research was sponsored by the Navy PEO C4I and conducted within the Forces and Resources Policy Center of the RAND National Defense Research Institute, a federally funded research and development center sponsored by the Office of the Secretary of Defense, the Joint Staff, the Unified Combatant Commands, the Navy, the Marine Corps, the defense agencies, and the defense Intelligence Community.

For more information on RAND's Forces and Resources Policy Center, contact the Director, James Hosek. He can be reached by email at James_Hosek@rand.org; by phone at 310-393-0411, extension 7183; or by mail at the RAND Corporation, 1776 Main Street, P.O. Box 2138, Santa Monica, California 90407-2138. More information about RAND is available at www.rand.org.

Contents

Figures

Tables

Summary

This study broadly assessed the manpower, personnel, and training implications associated with the introduction of the Consolidated Afloat Networks and Enterprise Services (CANES) to U.S. Navy ships. CANES will provide a common computing network and common operating system for command, control, communications, computers, and intelligence (C4I) systems onboard Navy ships, which could reduce the requirement for manpower and alter the demand for training. This environment will differ considerably from the traditional environment, which included stovepiped networks with unique hardware and software systems. The Navy effort to consolidate hardware and operating software and to introduce service-oriented architectures is consistent with the practices of private-sector organizations and information technology providers.

This RAND effort focused on particular Information Technology (IT) and Electronics Technician (ET) Navy Enlisted Classifications (NECs) associated with a subset of CANES systems, networks, and applications, including the Integrated Shipboard Network System (ISNS), the Sensitive Compartmented Information (SCI) Network, the Combined Enterprise Regional Information Exchange System–Maritime (CENTRIXS-M), the Global Command and Control System–Maritime (GCCS-M), and the Navy Tactical Command Support System (NTCSS). Given this selection from the list of CANES early adopters, this work focused primarily on the IT NECs specific to these systems. The analysis included two ship classes: carriers (CVNs) and destroyers (DDGs).

This report provides a review of current Navy manpower, personnel, and training practices; the implications of the conversion to CANES; and resulting recommendations.

Current Navy Practices

Manpower

This document describes the calculations used to develop the manpower requirements for IT and ET personnel. We determined IT requirements with a manpower equation that includes multiple inputs, such as Condition I (CI) watches, Condition III

(CIII) watches, the ship standard workweek, preventive maintenance, corrective maintenance, facilities maintenance, own unit support, and others.[1] Our analysis indicates that while IT manpower requirements on destroyers are determined by CI and CIII watches, CIII watches drive the IT manpower requirements on carriers. This is because the hours available for work beyond the watchstanding requirement exceed the hours needed for maintenance and other activities. Thus, despite technological improvements that would otherwise suggest reduced manpower requirements, no reduction of requirements is likely without a reduction in the watches. ET requirements are determined with a different requirements model: ET requirements are designated such that there is ET coverage for each type of equipment onboard.

Further, previous significant technology changes have had only very limited effects on manpower requirements. This observation underscores our conclusion that to affect manpower requirements, one must reduce watches, change organization, shift to another model for determining requirements, or eliminate equipment, in the case of ETs.

Manning

There are many manning or personnel issues pertaining to ITs onboard ships, but few of the issues are specific to CANES. For example, even if the requirements are exactly right, the ship's authorizations may reduce the manning on a ship. Another of the manning problems we discuss is that ships are detailed by aggregate NEC, without consideration for the number of people or whether the individual with that NEC will be available to the department in which the NEC is needed. Other problems mentioned reflect the traditional practices onboard ships that make junior IT personnel unavailable to their own departments for a significant portion of their initial assignment.

Training

There were several perceived deficiencies in IT training relayed to us from shipboard personnel, community managers, detailers, and training personnel. The first is the timing of the IT NEC training: Most ITs are assigned to their first unit without an NEC and have never actually touched the systems they will work with. This is in contrast to ETs, who primarily attend C school prior to their first assignment. Another issue is that the training software and hardware sometimes vary from the assignment destinations of the trainees. Shipboard personnel also note the difficulty of sending

---

[1]   CI is Battle Readiness, during which "[a]ll personnel are continuously alert" and "[a]ll possible operational systems are manned and operating. No maintenance is expected except that routinely associated with watchstanding and urgent repairs. Maximum expected crew endurance at Condition I is 24 continuous hours" (McGovern, 2005). CIII is Wartime Cruising Readiness, during which "[o]perational systems are manned and operating. . . . Accomplishment of all normal underway maintenance, support, and administrative functions is expected. Opportunity for eight hours of rest provided per man per day. Maximum expected crew endurance at Condition III is 60 continuous days" (McGovern, 2005).

personnel to training, for reasons such as pay grade prerequisites that exclude high-potential junior personnel, the NEC prerequisites that may leach their department of necessary skill capabilities, or other manning shortages. Still another aspect of training mentioned was the concern that training completion and award of the NEC do not necessarily reflect system expertise. This is sometimes a perceived issue of personnel completing sufficient training to pass the course but not enough to apply their knowledge to the challenging shipboard environment. This latter concern also reflected more senior personnel, whose training is not recent and whose system expertise has eroded.

Manpower, Personnel, and Training Implications for the CANES Program

Implications from the Literature

The literature indicates that manpower reductions from technological innovations are more likely if organizational and technological centralization exists. Moreover, IT insertions can facilitate structural and work redesign that leads to downsizing and increased productivity. For CANES, the implications from the literature are straightforward. Stakeholders, of which there are many in the Navy technology and manpower, personnel, and training enterprises, have a say in structural and work redesign. Neither organization nor technology decisionmaking is solely the province of the PEO and program managers. However, one should assume that technology insertions such as CANES should facilitate watchstanding changes and greater productivity; a smaller but more experienced IT workforce; fewer and less complex tasks; better training and tracking of NEC use and reuse; and the same fill but better fit of personnel to billets.

Manpower Implications for CANES

If we assume that IT manpower requirements continue to be determined by a watchstanding model, our analysis suggests that at least one CI watch could be eliminated from a destroyer. This reduction would equate to 6 percent of the IT manpower on a DDG. We also estimate that CIII watches on carriers could be reduced, for possible savings of 6 to 12 percent of IT manpower. Further, our analysis suggests that the ET 1678 NEC will likely not be needed for CANES, although this may not reduce the number of ET requirements, given that ET personnel tend to have more than one NEC and that the other may still be required.

Watchstanding is not the only basis for calculating manpower needs. An alternative model is a maintenance model in which IT workload is tied to own unit support and planned preventive and corrective maintenance. Another model is an engineering model in which unmanned spaces exist and equipment is centrally monitored via consoles and "rovers" are sent to the spaces as needed. Finally, a more experienced and better-trained IT workforce could lead to reduced requirements from improved productivity.

Manning Implications for CANES

Our analysis suggests several manning implications for CANES. First, the current detailing process of assigning personnel by aggregate NEC limits the effective use of CANES IT personnel; we recommend a more individualized assignment process for these more technical personnel. Our analysis also suggests that the traditional shipboard practice of using junior personnel away from their NEC for the initial year of their sea tour is a barrier to the effective use of IT personnel, especially if IT personnel were to receive additional training before their assignment. Additionally, converting the entire IT community to an initial six-year enlistment and providing C school before the initial assignment would be beneficial to CANES because of the resulting productivity gains. Longer initial enlistments may also result in long-term cost savings from assessing and training fewer IT personnel.

Training Implications for CANES

The Department of Defense (DoD) has issued DoD Directive 8570.01, which requires IT personnel, among others, to become certified in Information Assurance. This requirement will have positive implications for CANES, as it will ensure certain technical capabilities of those IT personnel working on CANES systems. Other training implications discussed in this report include plans to increase the length of IT A school, in part to accommodate the certification requirement; resequencing the IT NEC training; and moving C school to the beginning of the IT career. Although current Navy plans are to provide initial C school to only a minority portion of ITs, our analysis suggests the benefit of providing initial C school to all ITs, which would result in a considerable increase in the number of trained ITs assigned to units.

Recommendations

The first recommendation is specific to manpower, personnel, and training in the CANES environment. The next six affect all ITs and thus have significant implications for CANES. The last one affects many Navy ratings and is not a new suggestion.

- The PEO C4I should work with the Navy Manpower Analysis Center (NAVMAC) and with organizational stakeholders (e.g., the type commanders [TYCOMs]) to either reduce watches for ITs or move to a different model for addressing manpower requirements. Ideally, the manpower model selected would permit the Navy to capitalize on technology advances, such as those resulting in improved reliability and the opportunity for virtual administration, that would otherwise suggest a reduction in manpower.

- Proceed as planned with longer A school to provide Level One IA certification to IT personnel. However, also institute a two-week remedial program for those personnel who are not initially successful with certification.
- Add critical training elements from the 1678 NEC to IT network training to facilitate the absorption of the 1678 requirement among ITs.
- Consider greater use of the detailing strategy used on the Littoral Combat Ship (LCS). In other words, assign IT personnel as individuals to fill specific positions, and ensure that they receive appropriate training en route.
- Enlist all IT personnel with a six-year enlistment contract and send all ITs to C school following A school, in order to dramatically increase the number of trained ITs associated with CANES.
- Explore whether the early C school can reduce the length of system-specific NEC training. Additionally, if early C school is not instituted for all ITs, still consider resequencing NEC training such that network training is prerequisite for system-specific training.
- Consider whether the productivity gains from early C school should result in greater effectiveness or in manpower savings.
- Consider whether the traditional use of junior personnel onboard ships remains appropriate and effective, especially for highly trained technical personnel.

Acknowledgments

This study was sponsored by the Program Executive Officer for Command, Control, Communications, Computers, and Intelligence (PEO C4I). The PEO, Chris Miller, and our immediate study sponsors, Captain Joe Beel and Sean Zion, were extremely helpful to us, and we thank them for their guidance and support of the study. Additionally, throughout the course of the study, we had complete access to and support of other staff of the PEO and related organizations, such as Space and Naval Warfare Systems Command (SPAWAR), and we are grateful for their willingness to share information about the issues at hand.

Many other individuals contributed to the study. In particular, John Blayne, LCDR Scott Fairbank, ETCM Edward Ferber, LCDR Jamie Gateau, ITCS Stanley Greene, Gary Grice, Gregory Hayes, Mike Jones, Donald Kania, CDR Chris Lapacik, CDR Stephen Lorentzen, Harold Leupp, Wayne McGovern, LT Riley Murdock, ITCS Karl Parsons, Charles Sauter, CDR Craig Schauppner, CDR Abe Thompson, and Robert Wolborsky were extremely generous in contributing their time and sharing information about CANES and/or Navy manpower, personnel, and training processes.

We thank our RAND reviewers, Michael Hansen and Dan Gonzales, for their detailed and constructive reviews and helpful suggestions for improving the clarity and flow of the report. We also benefited from the editing of James Torr and additional input from RAND colleague Roland Yardley and from RAND's Navy fellow, CDR Daniel Cobian.

Abbreviations

ADNS	Automated Digital Network Service
ADP	automated data processing
AMD	activity manpower document
C4I	command, control, communications, computers, and intelligence
CANES	Consolidated Afloat Networks and Enterprise Services
CENTRIXS-M	Combined Enterprise Regional Information Exchange System–Maritime
CI	Condition I
CID	Center for Information Dominance
CIII	Condition III
CIP	Common Intelligence Picture
CMH	Corrective Maintenance Hours
CNP	Chief of Naval Personnel
COE	Common Operating Environment
COP	Common Operational Picture
COTS	commercial off-the-shelf
CS02	Information Resources Management Division
CS03	Resources Management Division
CS05	Data Division
CVN	carrier
DDG	destroyer
DoD	U.S. Department of Defense

DP	Data Processing; Data Processing Technician
EDVR	Enlisted Distribution Verification Report
ET	Electronics Technician
FMH	Facilities Maintenance Hours
GCCS-M	Global Command and Control System–Maritime
GOTS	government off-the-shelf
IA	Information Assurance
INE	Inline Network Encryption
ISNS	Integrated Shipboard Network System
IT	Information Technology; Information Systems Technician
LAN	local area network
LCS	Littoral Combat Ship
MH	Make Ready and Put Away Allowance Hours
MLTC	Multi-Level Thin Client
MRMS	Maintenance Resources Management System
MRPA	Make Ready and Put Away
N1	Deputy Chief of Naval Operations for Manpower, Personnel, Education and Training/Chief of Naval Personnel
NALCOMIS	Naval Aviation Logistics Command Management Information System
NAVMAC	Navy Manpower Analysis Center
NEC	Navy Enlisted Classification
NTCSS	Navy Tactical Command Support System
NTSP	Navy Training System Plan
OUSH	Own Unit Support Hours
PEO	Program Executive Officer
PH	Productivity Allowance Hours
PMH	Preventive Maintenance Hours
POE	Projected Operational Environment

PW	Productive Workweek
RM	Radiomen
ROC	Required Operational Capability
SMD	ship manpower document
SNAP	Shipboard Non-Tactical ADP Program
SCI	Sensitive Compartmented Information
SD	Service Diversion
SOA	service-oriented architecture
SPAWAR	Space and Naval Warfare Systems Command
SW	Ship Standard Workweek
TA	Training Allowance
TYCOM	type commander
UIC	unit identification code
WH	Watchstanding Hours
YOS	years of service

CHAPTER ONE

Introduction

The objective of this research was to assess broadly the manpower, personnel, and training implications associated with the introduction of a common computing environment—Consolidated Afloat Networks and Enterprise Services (CANES)—on naval ships. The sponsor of the research had an expectation that this technology could reduce the need for associated manpower but alter the demand for training. We identified the manpower, personnel, and training implications of the conversion from the current legacy networks, systems, and applications to understand the impact on the numbers and types of personnel needed and on overall training demands. Moreover, we reviewed personnel management policy for the Information Technology (IT) community to understand how it potentially affects the systems of the Program Executive Officer for Command, Control, Communications, Computers, and Intelligence (PEO C4I).

Background

The various C4I and warfare systems on naval ships and onshore installations are currently developed, fielded, and supported largely independently from each other. The result is numerous stovepiped networks, each with unique hardware and software systems. For example, the typical naval ship has more than 50 separate networks, each requiring properly trained support personnel. The Navy is starting to move toward consolidating hardware and operating software and introducing service-oriented architectures (SOAs) that offer the promise of providing increased C4I capabilities in a flexible and cost-effective environment. Open architectures decouple the hardware, networks, and software applications of current systems and provide a consolidated computing environment where the stovepiped systems and networks act as applications. This concept mirrors the current practices of private-sector organizations and information technology providers and promises a flexible, adaptable information environment.

The literature suggests that a common computing environment should have an impact on the numbers and types of personnel needed to administer, operate, and maintain C4I systems on naval ships. The consolidation should result in fewer categories and fewer specially trained personnel. This will affect both the manpower requirements for a ship's crew and the training infrastructure needed to provide the required skills. The Navy needs to understand the potential effects on manpower, personnel, and training requirements resulting from the implementation of CANES as well as of the changing personnel management environment required to support CANES.

Consolidated Afloat Networks and Enterprise Services

CANES is designed to function as the shipboard networking infrastructure resource that provides services to hosted applications. New hardware and operating software facilitate reductions in the hardware footprint and administrative and maintenance overhead while consolidating services. Systems such as the Navy Tactical Command Support System (NTCSS), the Combined Enterprise Regional Information Exchange System–Maritime (CENTRIXS-M), the Global Command and Control System–Maritime (GCCS-M), and the Sensitive Compartmented Information (SCI) Network Command and Control System are integrated into CANES. These systems have a significant network administration and maintenance overhead that shifts to the CANES administrator. About 20 other applications are currently hosted on the prototype Integrated Shipboard Network System (ISNS) Early Adopter Network preceding CANES, and even more applications will be hosted by CANES. CANES is designed to operate unattended, with network management tools continuously monitoring key system parameters and services.

What RAND Was Asked to Do

The RAND National Defense Research Institute study team was asked to work closely with the PEO C4I staff to identify appropriate legacy systems for analysis and to leverage existing analysis.

There are hundreds of different legacy systems on naval ships and onshore installations that could migrate to a common computing environment. The first task was to define the range of systems for further analysis. Iterating with the PEO C4I and the CANES program office, this task specified those legacy systems to include in the analyses. The list was drawn from those systems identified as early adopters of CANES. We also developed a list of ship classes to include in the analyses, as well as a list of Navy Enlisted Classifications (NECs) to be considered.

The next task was to develop a roadmap of the current numbers of personnel, the enlisted classifications, and training courses required to administer, operate, and maintain current systems on the selected naval ships. We were asked to interact with the organizations responsible for the legacy systems and the organizations that determine manpower and training requirements for a ship's crew.

The last task was to estimate the overall effect that the introduction of CANES would have on manpower, personnel, and training. Our goal was to specify the likely decrease, or increase, in the numbers and types of personnel for the ship classes of interest and the impact on the numbers of courses and student throughput dedicated to the training of CANES personnel.

Focus of Research

We focused our research on selected systems, ship classes, and NECs. The primary systems, networks, and applications of interest to the sponsor were ISNS, the SCI Network, CENTRIXS-M, GCCS-M, and NTCSS. These are the systems that largely drive the need for manpower and specialized training in the case of ITs. We examined two ship classes: carriers (CVNs) and destroyers (DDGs). For each ship class, we analyzed detailed information about manpower and personnel for one particular ship of the class as an exemplar. We looked at eight NECs, one in the ET rating and seven in the IT rating. Table 1.1 lists them.

Approach

We took both a top-down and a bottom-up approach. For the latter, we started by collecting essential information on each network and application and gathered views of the various stakeholders with respect to those networks and applications. That allowed us to establish a manpower, personnel, and training baseline for each of the systems. Working top-down, we reviewed research literature and case studies to discern potential manpower, personnel, and training implications. We then assessed significant inputs and intermediate outputs that could affect final manpower, personnel, and training outcomes with respects to CANES. Both the bottom-up and top-down approaches were useful for drawing conclusions and recommendations.

Organization of This Report

Chapter Two provides a review of Navy manpower, personnel, and training practices with respect to the IT rating and the NEC of interest to the study. Chapter Three provides our assessment of the manpower, personnel, and training implications for the CANES program. Chapter Four provides recommendations.

Table 1.1
Navy Enlisted Classifications of Interest

Designation	Name	Brief Description
ET 1678	Information Systems Maintenance Technician	Provides journeyman-level maintenance on shipboard information systems
IT 2710	Global Command and Control–Maritime 4 System Administrator	Performs installation, configuration, administration, repair, and basic operation of the system
IT 2720	Global Command and Control–Maritime System Administrator	Performs basic operation of the system with regard to the system administration functions
IT 2730	SNAP III System Administrator	Coordinates the implementation, operation, and software maintenance of the system and establishes and monitors security procedures
IT 2735	Information Systems Administrator	Administers commercial network operating systems, including configuration, system, and performance management and network software and hardware corrective action
IT 2779	Information System Security Manager	Serves as focal point and principal adviser for information security; analyzes and evaluates system security technology and policy; develops and maintains system accreditation and support documentation
IT 2780	Network Security Vulnerability Technician	Recognizes operating system vulnerabilities and performs corrective actions to ensure maximum system availability
IT 2781	Advanced Network Analyst	Manages network operating systems; implements connectivity solutions and protocols, services, and standards

NOTE: These designations for enlisted skills are in the process of changing as the Navy implements new personnel and training practices. The changes are reviewed later in the report. SNAP = Shipboard Non-Tactical ADP (Automated Data Processing) Program.

This report also includes four appendixes. The first provides descriptions of the CANES systems that were included within the scope of this work. Appendix B describes the simulation model used to analyze training alternatives, and Appendix C includes the equations for that model. Appendix D provides the more detailed analysis of community management and training options conducted with that model as well as the costing of those options.

Where the Navy Is (and Has Been)

This chapter reviews the evolution of manpower, personnel, and training for the selected systems and the occupational specialties that support them.

Information and Electronics Technician Ratings

Information Technology

The IT rating was established in 1999 through the merger of Radiomen (RM) and Data Processing (DP) ratings. (See Figure 2.1.) At the time, ratings mergers were encouraged by Chiefs of Naval Personnel (CNPs/N1s). More recently, however, the N1 has disapproved certain ratings mergers and slowed others for more study. As a result of these mergers, the current IT rating includes two relatively different kinds of person-

Figure 2.1
Evolution of the IT Rating

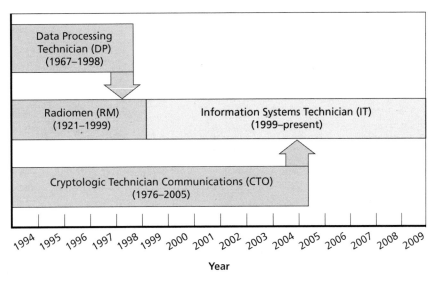

RAND *MG896-2.1*

nel: (1) those who work on radio-telephones and radio-teletypes, prepare messages, and are knowledgeable and/or responsible for portions of the antennae and satellite communication systems and (2) those who are knowledgeable of and/or responsible for the computer systems and network administration. While nothing precludes an individual from being trained across those two areas, the prerequisites required for more advanced training will generally focus an individual in either the communication or the computer aspects of IT. In practice, communicators and data processors have different NECs, are trained differently, and do not truly overlap until they become more senior.

There are about 10,300 ITs in the Navy. The rating is generally healthy; however, it is undermanned at the grades of E-1 to E-4. It is the second most shore-intensive rating in the Navy and recently had its first two sea tours reduced to 48 months from 60 months. About half of all ITs are on sea duty of one form or another at any given time. Within the IT rating, there are a number of NECs available. About 26 percent of ITs are "Quad Zero" (NEC 0000), which means that they have been through A school but not through a more intensive advanced training course to specialize. We were asked to look at seven particular IT NECs. Personnel with these NECs as their primary NEC account for 38 percent of all ITs, with the largest group being those with the Information Systems Administrator (2735) NEC. Of the NECs of interest to this study, 41 percent are on sea duty.

IT is an occupation that is changing. At a minimum, new entrants must be able to be Top Secret/SCI cleared. U.S. Department of Defense Directive (DoDD) 8570.1 (2004) and DoD Instruction (DoDI) 8500.2 (2003) have requirements for certification at three different levels, and the requirement is for all ITs to be certified at the level appropriate to their seniority. Although IT is not part of the advanced electronics career field, which has higher entry aptitude and initial terms-of-service requirements, making IT part of that field is under consideration and we discuss it later in the report.

Electronics Technician

Electronics Technician (ET) is a longstanding Navy enlisted rating, dating to 1948. Periodically, other ratings have merged into it. (See Figure 2.2.) ET personnel are responsible for Navy electronics equipment, which they maintain, repair, calibrate, tune, and adjust.

There are about 6,100 personnel in the surface portion of the ET rating. Of these, about 470 hold the NEC of 1678 that is of interest to this study. ET is also shore-intensive, with 57 percent on sea duty. In NEC 1678, 43 percent are on sea duty. The ET rating is part of the advanced electronics career field. All ETs enter the Navy for a six-year term of service and attend A and C schools prior to their first assignment.

Figure 2.2
Evolution of the ET Rating

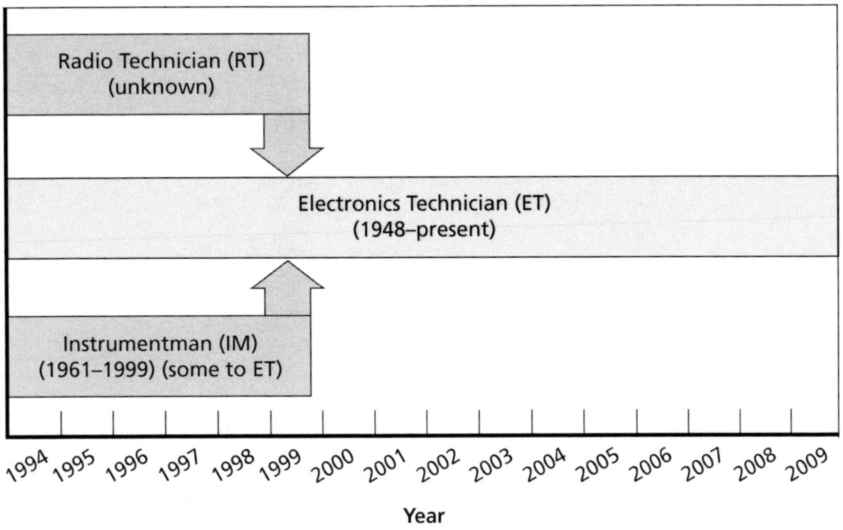

RAND *MG896-2.2*

Manpower and Training Practices

Information about billets and training required for certain systems is contained in Navy Training System Plans (NTSPs). These plans are required for new ships and equipment as well as for upgrades to equipment. Our review of these plans for the past ten years shows that consolidation of systems is not a new concept. For example, the ISNS NTSP states that ISNS "integrates network equipment, servers, client work-stations, and computer software into an open, scaleable, network centric architecture. . . . Traditionally, individual . . . programs have provided this capability via stovepipe, single-purpose LAN [local area network]."[1]

Similarly, NTCSS provided better capability than its predecessor systems:

> The NTCSS Program Office . . . has developed a series of hardware and software configurations that replaces very old, expensive to maintain, and unreliable equipment with new open-systems compliant equipment that is reliable and economical to own. This hardware modernization effort to Navy common PCs [personal computers] and Servers was approved in 1994 under the legacy programs of SNAP and NALCOMIS [Naval Aviation Logistics Command Management Information System]. NTCSS migrates software applications to a modern client-server/ RDBMS [relational database management system]/GUI [graphical user interface] environment and the GCCS-M Common Operating Environment (COE)

---

[1]   Department of the Navy, 2005. While this quote is from 2005, the original NTSP (N6-NTSP-E-70-0304) of August 2, 2004, had similar language.

for a single logistics support infrastructure and provides significantly improved performance, which enhances user productivity in all supported functional areas. NTCSS integrates together the three major command support programs (SNAP, NALCOMIS, and Maintenance Resources Management System (MRMS)) into one. (Department of the Navy, 2002)

However, the manpower requirement for IT personnel onboard ship to support the systems of interest has not evolved as the systems evolved. In the early days of technology insertion, there was concern about "computers" managing critical ship systems. The initial manpower logic was to establish a watch station wherever there was a server location. In essence, a human would be available 24/7 to deal with the equipment. And for the most part, this logic of manpower requirements tied to watch stations has not changed. As the PEO introduced newer systems, such as ISNS and NTCSS, no changes were made to the legacy practices of watchstanding for ITs. For example, although NTCSS consolidated several existing applications and systems, the summary manpower statement from the cited NTSP was that

> NTCSS . . . will not alter any current military duties, in port watches, or Condition I (General Quarters) assignments for operator or maintenance personnel. NTCSS operates on a 24-hour, 7-days-a-week schedule as needed in support of ship's work centers. The bulk of the system hardware is located in a central space and will be managed, as required, by NTCSS managers. A qualified NTCSS Manager must be available at all times to resolve software failures, restore system operation, and periodically monitor system operation. (Department of the Navy, 2002)

ISNS, which also integrated existing capabilities, stated, "The use of ISNS has no impact on watchstanding at various conditions of readiness. . . . The goal is to enhance watchstander efficiency within the current billeting structure" (Department of the Navy, 2005).

In essence, the original practice of requiring a watch wherever there was a server location was continued and, as seen in the next section, manpower requirements largely changed only as watch stations changed. In many respects, the use of IT onboard ship is more a product of legacy practices than of modern technology and organization designed to take advantage of that technology.

Derivation of Manpower Requirements: Ship Manpower Documents and Activity Manpower Documents

Manpower Process

In Navy parlance, *manpower* equates to demand—the need for people to staff ships, squadrons, and shore organizations. These needs take two forms. *Requirements* are the billets or spaces at the grade and occupational level of detail that are needed to perform the organization's mission. *Authorizations* are billets that have been funded either to the required grade and occupation or to less than that level. Shipboard authorizations never exceed requirements and are typically less than requirements. It is the authorizations that affect personnel distribution or manning. Requirements for a class of ships are stated in a ship manpower document (SMD). If a class of ships has multiple configurations, there is an SMD for each configuration. Requirements and authorizations for a particular ship are stated in an activity manpower document (AMD).

Both of these documents are the end result of the Navy manpower requirements process. The ship or fleet requirements themselves are derived from analysis that considers Required Operational Capability (ROC), Projected Operational Environment (POE), ship design, technology (new equipment), policy, ship/department/division watchstanding requirements, maintenance needs, and judgment. Policy can include mandated watches, Navy staffing standards, such as the length of the standard at-sea workweek, and directed requirements. Judgment includes results of on-site analysis and other reviews and expert opinion. Authorizations result from resource decisions with respect to the requirements and are typically set at less than the requirement.

The process is managed by the Navy Manpower Analysis Center (NAVMAC), and a number of stakeholders are involved in it. The detail of the process is automated and embedded in the Navy Manpower Requirements System at NAVMAC. New acquisition programs, such as CANES, have the potential to change ship manpower, manning, and training. An assessment of the impact of the acquisition is needed in order to revise NTSPs, which then form the basis of a revised SMD.

In order to conduct our independent assessment of manpower implications, we approximated mathematically the logic of the manpower requirements system, validated the results against historical SMDs, and then analyzed current manpower and CANES implications using the derived equations.

Mathematical Approximation of Manpower Requirements
Fifteen variables have significant effects on ship manpower requirements determination:

- CI: Condition I Watches (division, department, ship)
- CIII: Condition III Watches (division, department, ship)
- C3N: Condition III Watch-Standers = 3 x CIII
- SW: Ship Standard Workweek = 81hrs
- TA: Training Allowance = 7hrs
- SD: Service Diversion = 4hrs
- PW: Productive Workweek = (SW – TA – SD) = 70hrs
- WH: Watchstanding Hours = 56hrs
- PMH: Preventive Maintenance Hours
- CMH: Corrective Maintenance Hours
- FMH: Facilities Maintenance Hours
- OUSH: Own Unit Support Hours

- PH: Productivity Allowance Hours
- MH: Make Ready and Put Away (MRPA) Allowance Hours
- X: Other Hours

For the most part, two conditions of readiness govern manpower determination for ITs. Condition I, also known as general quarters, is a maximum state of readiness, and all assigned stations are manned for the duration of combat or the emergency. Condition III (deployed or wartime cruising) watches are key. The standard afloat workweek assumes that a unit is steaming in Condition III (wartime or deployed readiness condition) on a three-section watch, i.e., approximately one-third of needed manpower on watch each eight-hour period. As a result, CIII watches require three watchstanders for each watch.[2]

Policy decisions, such as the length of the standard workweek and the amount allocated to training and other needs (service diversion), also have effects. The Navy afloat workweek increased several years ago and currently allows for 81 hours, which consists of 70 productive work hours, 7 hours of training, and 4 hours of service diversion (e.g., inspections, sick call, and other administrative requirements). For a watchstander, 56 hours are allocated (8 hours and 7 days) to watches, with 14 hours weekly available for additional work.

Various types of maintenance needs are assessed through detailed analysis and policy decisions. Planned maintenance is routine preventive maintenance. Corrective maintenance is unscheduled work as a result of equipment malfunction. Facilities maintenance is that work needed to maintain cleanliness, sanitation, and preservation against deterioration. Own unit support is the duties needed to accomplish the ship's mission, such as resupply and administrative tasks. For the maintenance and support needs, allowance for productivity (e.g., difficult working condition, bathroom breaks) and to make ready and put away (set up and teardown times) are factored in. Policy decisions change these allowances from time to time—for example, by reducing the length of the productivity and make ready allowances.

Two computations, one for hours available and one for hours needed, are necessary for the calculation of manpower needs:

$$\text{Hours available for work (H)} = [(PW + 0.9)*\text{Max}(CI \text{ or } C3N)] - WH*C3N$$

$$\text{Workload hours needed (N)} = PMH + CMH + OUSH + FMH + PH + MH + X.$$

---

[2]   CI is Battle Readiness, during which "[a]ll personnel are continuously alert" and "[a]ll possible operational systems are manned and operating. No maintenance is expected except that routinely associated with watchstanding and urgent repairs. Maximum expected crew endurance at Condition I is 24 continuous hours" (McGovern 2005). CIII is Wartime Cruising Readiness, during which "[o]perational systems are manned and operating. . . . Accomplishment of all normal underway maintenance, support, and administrative functions is expected. Opportunity for eight hours of rest provided per man per day. Maximum expected crew endurance at Condition III is 60 continuous days" (McGovern 2005).

Hours available for work is the productive workweek of 70 hours (plus a small fudge factor) multiplied by the maximum of CI or CIII watchstanders (three for each CIII watch) less the 56 hours that CIII watchstanders are each standing watch. Workload hours needed is the sum of the maintenance needs exclusive of allowances.

The manpower requirement, billets needed, is then determined by logic:

$$\text{If } H \geq N = \text{Max(CI or 3*CIII)}$$

$$\text{If } H < N = \text{Max(CI or 3*CIII)} + \text{Roundup } [(N - H)/PW].$$

If hours available for work beyond watchstanding exceed hours needed, the requirement is the maximum of CI, or three times CIII watches. If hours available are less than hours needed, then the additional work hours divided by the workweek length and rounded up is the number of additional billets to add to the watchstanding requirement. On-site reviews may lead to adjustments to these calculations.

Finally, *authorizations* is the number and quality of the billets that can be resourced. This is a resource, not a manpower decision.

Past Manpower Requirements for DDGs and CVNs

We used the equation and logic derived above in conjunction with historical SMDs to both validate our arithmetic and understand how requirements for IT personnel in the NECs of interest to us had changed over the years. The ship manpower requirement includes the watchstanding requirements and the estimates of maintenance workload. During the period for which we had data, new technology (equipment) for the applications/networks of interest had been introduced into the fleet, as discussed earlier.

Changes in CVN Division IT Manpower Have Been Watch- and Policy-Driven

On a Nimitz-class carrier, the combat systems department has several divisions. Two of them, Information Resources Management (CS02) and Resources Management (CS03), are composed of IT personnel in the NECs of interest to us. As shown in Figure 2.3, there have been changes to the requirements for these two divisions between 1999 and 2008.

Starting from the left with the Information Resources Management Division, in 1999, the requirement resulted from 6 CIII watches and additional workload to support one billet, for a total of 19 requirements. In 2004, two additional CIII watches and policy changes affecting the workweek and allowances led to a requirement of 24, all based on the 8 CIII watches. The 2008 AMD mirrors this requirement, but a Nimitz-class carrier is authorized two fewer billets than the requirement.

For the Resources Management Division, the requirement decreased between 1999 and 2004 as watch requirements and policy changed. The requirement for 10 bil-

Figure 2.3
Change in CVN IT Manpower

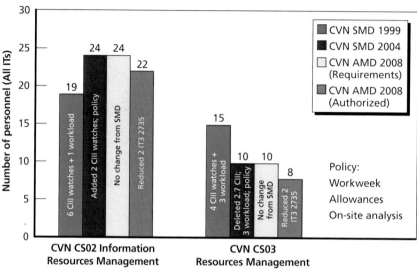

RAND *MG896-2.3*

lets is a combination of watches, additional workload, and on-site analysis that added 3 billets. The AMD for 2008 has a requirement for 10 billets, but the ship is authorized two fewer with NEC 2735.

For both divisions, requirements have changed over time, but those changes were the result of CIII watchstanding changes and policy (workweek) changes. There are sufficient hours after watchstanding to accommodate the maintenance needs, so maintenance is not a factor in manpower requirements for IT billets.

Figure 2.4 is based on the same data, but shows the information for the ITs with NECs of interest rather than organizationally.

In 1999, Radiomen (RM) and Data Processing (DP) had not yet merged into the IT community, so we analyzed the sum of the two in the figure. A Nimitz-class carrier required a total of 10 CIII watches (30 requirements) and 4 additional for maintenance workload, for a total of 34 requirements. By 2004, the CIII watches had increased to 40 requirements, with a reduction of workload requirements to 3. As stated above, an on-site analysis by NAVMAC added 3 additional requirements, for a total of 46. As of 2008, the AMD has a requirement for 46 billets, but is authorized 4 fewer 2735s than the requirement, for a total of 42.[3] We also looked at ET 1678s that are part of the Data Division (CS05). ET 1678s stood no watches in 1999 but had 4 requirements based

---

[3]   We will discuss the impact of authorizations being less than requirements later in the report. However, when authorizations are less than requirements, it makes it more difficult to claim an actual manpower savings from reducing requirements until requirements go below the level of authorizations. Moreover, anecdotally, most assert that the Navy system will continue to lower authorizations if requirements are reduced.

Figure 2.4
CVN Selected IT and ET

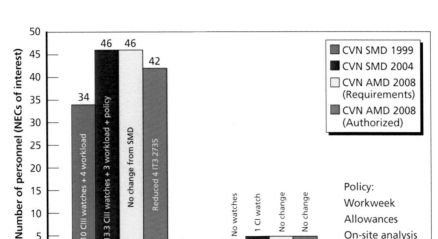

RAND *MG896-2.4*

on the necessity to have billets with equipment knowledge distributed in the ship. So ET is a special case of requirements, where the requirement is based primarily not on watches or on maintenance but rather on the need to have one or more billets with knowledge of particular equipment onboard ship. For critical equipment, the need is at least two billets (either primary or secondary NEC) with more needed on larger ships. One CI watch was added in 2004, and the 2008 requirement and authorization is for five billets.

DDG IT Manpower Has Changed Through Mergers and Watches

DDGs have several configurations. Flight I is significantly different from Flights II and IIA, which are more similar to each other. Figure 2.5 shows the IT requirements changes over time for both configurations. In the 2003–2004 timeframe, two IT manpower requirements were removed. One removal resulted from a merger of two divisions; the other resulted from policy changes and change in the workweek. In 2006, one IT requirement was added to Flights II and IIA for an additional CIII watch. In 2000, one ET with a secondary NEC of 1678 was added to Flight II. Both flights of the DDG currently have one ET with a primary NEC of 1678 and one with a secondary NEC of 1678.

Current Manpower Requirements for DDG and CVN

CVN and DDG manpower requirements are both driven by watchstanding, but each is based on different watchstanding readiness conditions.

Figure 2.5
DDG IT Manpower

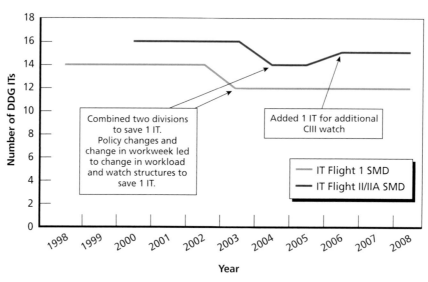

Combined two divisions
to save 1 IT.
Policy changes and
change in workweek led
to change in workload
and watch structures to
save 1 IT.

Added 1 IT for additional
CIII watch

— IT Flight 1 SMD
— IT Flight II/IIA SMD

Number of DDG ITs

Year

RAND *MG896-2.5*

DDG IT Manpower Requirements Result from Condition I and Condition III Watches

Table 2.1 shows watch requirements for the NECs of interest to us. As calculated from the manpower equation, CI watches require one billet each, while CIII watches require three billets each (three eight-hour shifts).

Table 2.1
DDG Manpower Requirement (NECs of Interest)

Watch Stations/Titles	NEC	Need in Condition I	Need in Condition III
Ship's Signals Exploit Space			
SCI Networks Supervisor	2781	IT1	
SCI Networks Operator	2735	IT3	ITSN
Information Systems			
Network Security Technician	2780	IT1	
NTCSS System Administrator	2730	IT2	
LAN Manager	2735	IT2	IT3
Tactical Systems Administrator	2720	IT2	
Total Requirements		**6**	**6**

CVN IT Manpower Requirements Result from Condition III Watches

Table 2.2 shows watch requirements for the NECs of interest to us. As calculated from the manpower equation, CI watches require one billet each, while CIII watches require three billets each (three eight-hour shifts). A significant difference between the DDG

Table 2.2
CVN Manpower Requirement (NECs of Interest)

Division	Watch Stations/Titles	NEC	Need in Condition I	Need in Condition III
	Technical Control			
CS01	Operator #6	2735	IT3	IT3
	Network Control Center			
CS02	Computer Control Area Supervisor	2781	IT1	IT1
CS02	Information Systems Technician	2735	IT2	IT2
	Information Security			
CS02	Information Systems Supervisor	2779	ITC	
CS02	Network Security Technician	2780	IT1	IT2
CS02	SIPRNET System Administrator	2735	IT2	IT2
CS02	NIPRNET System Administrator	2735	IT2	IT3
CS02	HM&E System Administrator	2735	IT3	IT3
CS03	NTCSS System Administrator	2730	IT1	IT2
CS02	IDS Manager	2780	IT2	IT2
	CVIC Automated Data Processing			
CS05	GCCS-M System Administrator	2720	IT1	IT2
CS05	GCCS-M System Administrator	2720	IT2	IT3
CS05	GCCS-M System Administrator	2720	IT3	IT3
CS02	NTDIS Administrator	2735	IT3	IT3
	Division Allocated Watches			
CS03	Damage Control, Utilityman	Any	IT	
CS03	Damage Control, Nozzleman	Any	IT	
CS03	Damage Control, Talker	Any	IT	
CS03	Weapons Control, Alarm Monitor	Any		IT2 (1 shift)
Total Watches			**17**	**13.3**
Total Requirements			**17**	**40**

NOTES: SIPRNET = Secure Internet Protocol Router Network; NIPRNET = Nonsecure Internet Protocol Router Network; HM&E = hull, mechanical, and electrical; IDS = Integrated Display System; CVIC = Carrier Intelligence Center; NTDIS = Navy Tactical Data Information System.

and the CVN is that, on the CVN, the bulk of the manpower requirement is driven by CIII watches.

Summary

Significant technology shifts in the past have had limited manpower effects by themselves. To have an effect, one must reduce watches (for ITs), change organization (merge roles), or eliminate equipment (for ETs). This means, in sum, that technological changes that would suggest reduced manpower needs—e.g., because of increased reliability or the opportunity to administer and otherwise support the systems virtually from shore—will not result in reduced manpower as long as the Navy continues to use a watchstation-based manpower requirements calculation.

This becomes more apparent when considering specific types of ships. The number of ITs on smaller ships is as much driven by CI watches as by CIII watches. Eighty to 100 percent of DDG crew have CI watches.[4] Thus, no reduction of requirements is likely without changing watchstanding needs at CI. The number of ITs on larger ships tends to be driven more by CIII watches. Only about 66 percent of a CVN crew have CI watches, and thus there is more leeway to reduce on a CVN by making changes to CIII watches.

There is a potential opportunity to reduce ET manpower, as ET 1678s may not be needed in the CANES environment as the equipment becomes more plug-and-play and requires replacement rather than repair. If tasks remain, they could be absorbed by the IT personnel onboard.

In the next chapter, we offer specific manpower implications for the CANES environment.

Personnel Issues: IT and ET Manning

Many of the personnel issues that were discussed during interviews onboard ships and with the community detailers were not specific to IT and ET personnel. Nonetheless, we provide some manning observations regarding IT and ET personnel. In short, even perfect requirements do not lead to perfect manning. Less-than-perfect manning is an outcome of whether requirements are authorized, whether there is sufficient inventory that can be assigned to authorized billets, the detailing process, and the availability of people once assigned.

---

[4]   In practice, all sailors onboard may be assigned a CI watch. However, for requirements purposes, NAVMAC counts only those that are required.

Community Management

Community managers are responsible for the community-specific policy and decisions regarding accessions, retention, and overall management. In the case of ITs and ETs, community managers report that the overall numbers suggest that these communities are relatively healthy. The inventory of ITs satisfies 97 percent of the FY2009 authorizations,[5] and ET inventory equates to 91.5 percent of authorizations.[6] There are some shortages among the most junior ITs because the merging of the CTO and IT communities resulted in a requirement for ITs to have Top Secret/SCI clearances. This security requirement has been associated with the higher-than-usual attrition of new recruits, but the Navy has begun to conduct some basic security-relevant assessments, such as of financial status and parental citizenship, before the recruits arrive at the Military Entrance Processing Station (MEPS), and the community management reports that the effort has helped IT accessions.

One of the community management issues reflects the inclusion of ITs in the recent Top Six Alignment. In 2000, the Navy planned an increase in the seniority of the enlisted billet force "reflecting the demand for more technical and experienced personnel required to operate the fleet's technologically advanced platforms and systems."[7] While this would appear to be the right direction for communities such as ITs, the Navy has recently reversed this increase for a number of communities, including ITs, in order to address the budget mismatch between authorized billets and the actual dollars allocated for the enlisted force. This reversal has been named the Top Six Alignment. As part of the Top Six Alignment, IT billets on ships have been decreased in pay grade. For example, on a carrier, one senior chief billet became a chief petty officer billet, one IT1 billet became an IT2 billet, three IT2s became IT3s, and five IT3 billets became ITSN billets. On a destroyer, one IT1 became an IT2, and two IT3 billets became ITSN billets. We conducted our interviews before this realignment, and the lack of sea duty experience among ITs was already a topic of concern among shipboard personnel. The decreases in paygrades resulting from the Top Six Alignment decreases further the likelihood that ITs assigned to ships will have system or prior sea duty experience. Further, the Top Six Alignment emphasis can be contrasted directly with the emphasis of the manpower and manning plans for the Littoral Combat Ship (LCS), which is designed as an optimally manned ship. Because there is no additional manning on the LCS, "the need to be cross trained and skilled to perform their jobs with minimal assistance or instruction is paramount and dictates a more experienced and senior crew" (Naval Sea Systems Command, 2007).

---

[5] January 2009 data.

[6] Communication with community managers, November 2008.

[7] Top Six Alignment Deck Plate Explanation, documentation provided by N12, Total Force Requirements Division, Bureau of Naval Personnel.

Detailing: Use of Enlisted Distribution Verification Reports and Multiple NECs

An Enlisted Distribution Verification Report (EDVR) is maintained for each ship (or shore) unit identification code (UIC). This document provides a monthly update of the present and known future manning status; it lists all enlisted personnel assigned to an activity, such as a ship, and the NECs that they hold (up to five). The Navy detailers use the EDVR to identify the manning requirements on ships. However, the detailing process proceeds by aggregate NEC. In other words, personnel are assigned to ships by NEC, not by the number of people. As a result, if an individual has multiple NECs, the single individual can satisfy a ship's need for multiple NECs. For example, a ship that needs a 2735 and a 2781 will likely receive one person, as the individual holding 2781 will also have 2735, since it is a prerequisite NEC for 2781. There are a couple of concerns about the practice of assigning by NEC. First is that the need for multiple NECs may sometimes convey the need for multiple people, but the detailing process will aim to send the fewest individuals to satisfy the NECs required. Second, the NECs are counted by ship. As a result, there may be individuals with the needed NECs onboard the ship, but they are not available to do the work for each NEC. For example, if the individual with a particular NEC has a leadership role on the ship, he or she may not be available to the division for hands-on work, but the individual will still count as satisfying the ship's need for that NEC.

These detailing challenges are recognized in the Navy's push toward "fit" rather than "fill."[8] *Fill* refers to placing an individual in a job, whereas *fit* implies identifying and satisfying the knowledge, skills, abilities, and tools required for a position. In detailing, *fit* refers to assigning an individual with the correct pay grade, rating, and NEC. IT detailers aim for 70 to 80 percent fit.[9]

In the aggregate, a carrier had somewhat less than 70 percent fit in a recent month, while a destroyer had 65 percent fit at the NEC level of detail.

Use on Ship and Availability for Work

In interviews, personnel on ships indicated several reasons that they do not have full use of all personnel assigned to them, according to the EDVR. When junior personnel arrive for their first tour at sea, they typically spend the first 12 to 18 months of that assignment performing general ship duties, such as working in the mess hall. Additionally, because ITs generally receive only A school before their first assignment, and because A school is based on computer-based training that does not provide hands-on experience with the actual equipment, a new IT initially needs to shadow a more experienced IT for on-the-job training. Reflecting the time lost to other ship duties as well as the time needed to gain expertise, carrier personnel estimated that they have real use of ITs for half of their assigned time. Destroyer personnel made a similar statement:

---

[8]   See, for example, Hoewing, 2004.

[9]   Interview with IT detailing personnel.

They estimated they have use of one-third of their assigned personnel at any one time. In addition to the reasons already stated, personnel are also lost to the department if they are TAD (assigned to temporary additional duty), such as for training; if they are nondeployable for health or other reasons; or if they are temporarily detailed away from the ship as individuals to support military operations.

Summary

In short, many factors contribute to personnel issues with ITs aboard ships, but few of them are specific to CANES, although all have the potential to impact CANES. Instead, they reflect the authorized billets and shortcomings in the current detailing processes for all ratings as well as the traditional practice aboard ship, such as the sharing of the general ship's duties across all junior personnel, regardless of their capabilities or the investment made in training those junior personnel.

Training

There are several perceived training deficiencies that were relayed to us during our interviews with community managers, detailers, trainers, and shipboard personnel. These generally pertain to ITs rather than to ETs.

First is the timing of the IT NEC training. Currently, most ITs attend A school and then are assigned to units as Quad Zeros, without an NEC. Further, the A school training that they do receive is based on computer simulations of the equipment. As a result, shipboard personnel complain that ITs do not have sufficient training to be effective when they arrive. Figure 2.6 indicates the current progression of, and prerequisites for, NEC training applicable to CANES. As the diagram demonstrates, while ITs may receive system-specific training, such as that for NEC 2720 or 2730, they do not receive networking training before their first assignment. This is in contrast to ETs, who attend C school prior to their first assignment, and this contrast was often pointed out to us by shipboard personnel as they compared the well-trained junior ETs with junior ITs.

Another concern regarding the training is that the training equipment or software at the training location is sometimes out of sync with the shipboard environment; either the ship or the training site may have more recent upgrades. This is a difficult shortcoming to address, and one that, were it not for the complaints regarding computer-simulation-based training, would suggest greater use of such training, since simulations can be more easily upgraded than hardware systems. Frequently, the comments regarding computer-simulation-based training were expressed as concerns about the effectiveness of such training, given that personnel do not actually "touch" real equipment. We heard, for example, that A school graduates arrive at their first ship

Figure 2.6
IT Training Courses of Interest

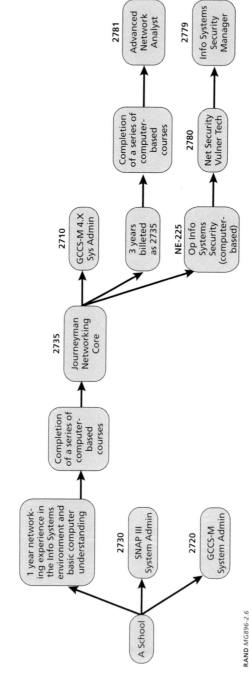

RAND MG896-2.6

assignment without ever having actually seen the equipment with which they will be working.[10]

Shipboard personnel also cited difficulty in sending personnel to training, for several reasons. The manning issues cited previously exacerbate the difficulty in releasing an individual to train. Also, the training often has pay grade prerequisites that preclude the ship from sending exceptionally bright, but junior, personnel. Additionally, because prerequisites are necessary for more advanced NECs, sending an IT to receive 2781 training, for example, may reduce the number of 2735s onboard to a critical level.

Another aspect of training that was mentioned was that completion of the course, and award of the NEC, does not necessarily indicate mastery or sufficient expertise. This reflects two concerns or observations. First, there are ITs who hold an NEC but whose training was sufficiently long ago that much of their knowledge has expired, especially if they have filled assignments that did not require use of that expertise. Second, there is concern that the training process does not rigorously test ITs before assigning an NEC. Those who held this perception sometimes asserted that some ITs learn enough to complete the training but not enough to apply their knowledge in a shipboard environment.

Summary

Training deficiencies discussed regarding ITs include the lack of hands-on training opportunities for new ITs; that NEC training occurs relatively late for ITs, who typically reach their first assignment without an NEC; that the training equipment or software is sometimes out of sync with that found in the shipboard environment; the difficulty of sending shipboard personnel to training; and that award of an NEC may not indicate mastery or advanced expertise.

The next chapter looks forward and addresses the manpower, personnel, and training implications for the CANES program.

---

[10] Subsequent to our interviews, a Navy Inspector General study identified problems that result from training recruits on computers, including offering sailors little hands-on experience. See "Computer-Based Failure," 2009.

Manpower, Personnel, and Training Implications for the CANES Program

Based on our review of manpower, personnel, and training processes and issues, this chapter discusses the implications for the CANES program. The chapter is based on prior research; data, including ship and activity manpower documents, Navy Training Systems Plans, Enlisted Distribution Verification Reports, watch bills for particular ships, and Navy and Defense Manpower Data Center personnel inventories and authorizations; interviews with personnel from the Space and Naval Warfare Systems Command (SPAWAR), the staff of the N1, SEA 21, the Navy Personnel Center, the Navy Manpower Analysis Center, the Center for Information Dominance; Navy uniformed and civilian personnel and subject-matter experts; and ship visits. We also use data analysis and system dynamics modeling to draw our observations.

Relevant Literature and Studies

We reviewed selected theoretical and empirical literature that was germane to this study, and we provide short summaries of the literature below. We also reviewed selected Navy-specific studies. We draw implications for CANES from this literature.

Empirical

Pinsonneault and Kraemer[1] reviewed a number of empirical studies that examined the impact of information technology on the number of middle managers in organizations and drew conclusions from them. The effect of information technology insertions on the number of managers is contingent upon the degree of centralization of computing decisions and organizational decisions. If both are decentralized, the number goes up; if both are centralized, the number goes down. If the two are not congruent, information technology has little effect. Pinsonneault and Kraemer conclude that information technology facilitates organizational downsizing but does not cause it. Information

---

[1]   Pinsonneault and Kraemer, 2002. See also Pinsonneault and Kraemer, 1997.

technology can facilitate structural and work redesign that leads to downsizing, but its effect depends on the context in which technology is used and on how it is used.

Bresnahan, Brynjolfsson, and Hitt (2002) conclude that information technology is a source of increased demand for skilled labor and is generally less costly to implement than organizational redesign that may also have a large effect on skill demand. "As information technology grows cheaper and more powerful, it induces more and more complementary investment in the rest of the cluster of changes—most importantly . . . in skilled labor." Their data show that skilled labor increases with three distinct changes at the firm level: information technology, new work organization, and new products and services. While new information technology can be easily implemented, organizational changes are often difficult, costly, and uncertain.

Stymne, Löwstedt, and Fleenor (1986) make two observations. If new information technology increases the level of productivity, then administrative and information handling employment can decrease. However, the effects of technology insertion cannot be understood without reference to organizational peculiarities and regulating mechanisms. The effects of information technology change will always be mediated and regulated by organizations through "buffers and absorbers."

> At least for some time, employment may increase during the change process: new specialists are needed to handle the new technology, people are needed for making studies and sitting in meetings, political infighting consumes resources, trainers are needed to teach the old employees the new system, alternates are needed to run the business when ordinary jobholders are away for training, and additional labor may even be needed to clear up the mess created by the switch-over.

The characteristics of decisionmaking processes and other organizational factors mitigate the effect of technological change. "Motivators and multipliers" are needed to gain less employment from information technology insertions.

Theoretical

Chan (2000) states that information technology can assume any of three roles—initiator, facilitator, or enabler—depending on the business environment and how the technology is being applied. Information technology promotes a business environment that is more efficient, more adaptable, and more flexible at all levels. However, the human elements (e.g., personality and culture) play major roles in organizational operations, including the effective and efficient deployment of information technology. Technological advances endow workers with an increased sense of control and degree of autonomy and heightened skill levels.

Gali (1999), in a macroeconomic study, points out that positive technology shocks decrease hours worked in the short term and then plateau off; productivity steeply increases in the short term and then levels off. Chang and Hong (2003) challenge this finding and conclude that while some industries do exhibit temporary reduction in hours in response to a permanent increase in total productivity, there are far more industries in which technological progress significantly increases hours.

Navy-Specific

There have been studies specifically about the Navy as well. Moore et al. (2002) state that for the Navy to reduce crew sizes without sacrificing readiness, other resources must be substituted. Choices include workload-reducing technology, using more skilled and experienced sailors, and using crew members more efficiently by eliminating unnecessary work, manipulating work schedules, or cross-training. Moore et al. state that the mathematics of billet creation limits the realization of billet savings from technology:

> Of course the long life of ships limits the range of alternatives; today's manpower planners must work with design decisions that may have been made decades ago. Still, part of the problem rests with business practices of today; an absence of incentives, organizational stovepiping that separates technology and manpower decisions.

The authors conclude that "to take full advantage of the manpower-reducing effects of technology, it may be necessary to reorganize work or employ more skilled people." Moore also cites other studies that have reached similar conclusions:

> Sims [1997] concluded that reorganizing work schedules and cutting watch-standing requirements would entail greater manning reductions than installing information technology. "However, new technology may indirectly affect manning by providing a rationale for watch-standing reductions that could have been made anyway." According to [Klein, Militello, and Crandall, 2000], new technology helps only "when viewed as part of a larger re-thinking of the organization."

Koopman and Golding (1999) state that the general conclusion by studies examining the relationship between technological advances and workforce skill levels is that "as technology gets more advanced the workforce becomes more rather than less skilled." Moreover,

> trends in both information systems and maintenance indicate that in the future, operator/decisionmakers will replace specialized maintenance technicians. . . . With added redundancy and reliability, there may be less need for the crew to know how the machines work and be able to maintain and repair them while deployed.

Stoloff et al. (2006) reviewed the utilization of personnel with NECs in the Navy. Eighty-two percent of new awards of an NEC are used at some time. Most use is first assignment after C school (70 percent). About 11 percent are never used before the individual separates from the Navy. Only 37 percent of NECs are reused at least once.

There is a significant benefit of NEC reuse from cost avoidance of initial NEC training. The IT rating avoids approximately $15,000 training cost per reuse. "The more significant benefit of NEC reuse is the cost avoidance associated with having to train new sailors, and the accumulation of human capital associated with keeping experienced sailors on the job." However, the current IT system does not directly support tracking and optimization of NEC utilization.

Garcia, Gasch, and Wertheim (2002) analyzed the information technology workforce to understand the work performed by this workforce and assess options for enhancing its training and professional development. Their findings include that the information technology A schools "fail to cover 79 and 60 percent of mission critical tasks in Information systems Administration and Communications, respectively." Also, 62 percent of E-4 and below reported in their survey that their A schools were not useful or only slightly useful. Not a single officer or senior enlisted recommended leaving the IT A school as is. Enhancing IT training through any one of three options analyzed, including redesigning IT A and C schools, is cost-effective. Among the benefits are avoided on-the-job training (2.5 hours per week for direct training of each recent graduate), avoided repair workload (8 hours per week because new System Administrators and Communicators were not adequately trained), avoided site visits by technical representatives, and avoided later schoolhouse cost to train. These benefits were larger than the cost of providing up front C school training. They recommend enhancing information technology training to cover all mission-critical tasks and integrating the LAN Administration C school into A school. Under this latter option, an additional 1,100 IT A school graduates would go on to LAN Administration training per year.

Summary of Literature Review
The manpower- and personnel-related effects of technology insertion are as follows:

- Reductions are possible if organizational and technological centralization exists.
- The effects of IT innovations cannot be understood without reference to organizational peculiarities and regulating mechanisms.
- IT innovations facilitate structural and work redesign, which leads to downsizing.
- IT improvements induce more investment in skilled labor.
- IT improvements increase productivity and reduce hours worked.
- Characteristics of the decisionmaking process and other organizational factors mitigate the effect of technological change.
- Technological advances endow workers with increased autonomy and heightened skill levels.
- Organizational stovepiping separates technology and manpower decisions.

- New technology provides a rationale for watchstanding reductions that could have been made anyway.

And the training-related effects of technology insertion are as follows:

- Restructuring jobs into job families simplifies training and eliminates redundant training.
- Significant benefit of NEC reuse from cost avoidance of initial NEC training.
- Enhancing IT training is cost-effective.

The implications for CANES from the literature are straightforward. Stakeholders, of which there are many in the Navy technology and manpower, personnel, and training enterprises, have a say in structural and work redesign. Neither organization nor technology decisionmaking is solely the province of the PEO and program managers. However, one should assume that technology insertions such as CANES should facilitate watchstanding changes and greater productivity; a smaller but more experienced IT workforce; fewer and less complex tasks; better training and tracking of NEC use and reuse; and same fill but better fit of personnel to billets.

The next sections provide implications for CANES based on our analysis and assessment.

Manpower Implications for CANES

Potential Watchstanding Changes, Given the Watchstanding Model

We first assume that the current watchstanding model used by the Navy Manpower Requirements System for IT does not necessarily change with the introduction of CANES.[2] There are a total of 15 IT requirements for a DDG, of which 13 require an NEC. Six of these requirements hold the NEC of interest to this study. Our limited interviews and data collection suggest that ships do not currently use personnel as the SMD would suggest, so change is feasible. For the DDG, watchstanders provide enough additional hours to meet maintenance needs now and projected into the CANES environment when they are forecasted to be less.[3] If CIII watches on a DDG for the NECs of interest could be reduced from two to one (six people to three), the manpower requirement would not change unless the six CI watches are also reduced. Currently, six requirements are needed for CI and CIII. (Three watchstanders are needed for each CIII watch; one is needed for each CI watch.) For each CI watch reduced, the manpower requirement is reduced by one requirement until the

---

next "step" of the remaining CIII watch is reached. Our assessment is that at least one CI watch could be eliminated. This amounts to about 6 percent of IT manpower on a DDG. Reducing another CI watch would be a 12 percent savings. Additional manpower savings are also possible with changes in NEC and training, as was suggested earlier.

For the CVN, IT watchstanders in most areas also provide enough additional hours to meet maintenance needs. In contrast to the DDG, where CI watches would have to be reduced as well as CIII, a reduction of CIII watches on a CVN leads to a reduction in three manpower requirements. This could take place in either the CS02 or CS05 divisions. Besides watchstanding requirements, CS03 division has three requirements based on on-site review and analysis and three based on workload, and this workload should be reduced in the CANES environment. One CIII watch is 7 percent of the CVN IT manpower we analyzed. Reducing the workload requirement to one provides another savings of two requirements. Division mergers could also lead to manpower savings. Changes to NEC and training of the type to be discussed also could have comparatively large effects on CVN manpower.

While ET manpower savings are possible if 1678s are no longer required, this may not necessarily reduce overall ship manpower, as ETs tend to have more than one NEC and the other may still be required.

Thus, even with maintaining the traditional watchstanding model for assessing IT manpower requirements, manpower savings of 6–12 percent of IT manpower appear feasible.

Other Manpower Models

Watchstanding is not the only basis for calculating manpower models, and this might be the juncture to begin to move away from this model for IT requirements. As stated earlier, the watchstanding model results from the earliest insertion of computers on ship. One potential model is to move to a maintenance model, in which IT workload is tied to own unit support and planned preventive and corrective maintenance. This is the other half of the manpower calculation reviewed previously. SPAWAR has moved in this direction by assessing the maintenance needs for the CANES environment. Another possible model is that used by the ET community, in which the number of IT onboard would be tied to the need to manage the CANES equipment and not to watchstanding. Still another model is one that could be called the engineering model, in which unmanned spaces have become more the norm. Equipment is centrally monitored via consoles, and "rovers" are sent to the spaces as needed and are not expected to be in those spaces 24/7. The Navy already has defined processes for managing this. Finally, a more experienced and better-trained IT workforce could lead to reduced requirements simply from improved productivity.

CANES is being tested on the Lincoln strike group, and it is possible to test and gather data about manpower requirements from that experiment. New approaches to manpower data analysis used by the DDG 1000, the LCS, and the LPD (Landing Platform Dock) 17 are also plausible. For example, the Total Crew Model is a discrete simulation that analyzes all tasks to be performed and ensures sufficient manpower to do them.

Personnel Implications

Manning to Specific Crew Positions

The current manning practices suggest that one of the greatest limitations to effective use of CANES IT personnel aboard ships is the detailing practice of assigning by aggregate NEC. In this regard, the LCS planned manning sets a useful precedent for CANES. For the LCS, the plans indicate that enlisted personnel will be detailed to the LCS similarly to the detailing process for officers. In this model, an individual sailor with appropriate prerequisite skills will be selected for and assigned to a specific crew position on the LCS. Further, this sailor will complete the billet specialty training en route to the LCS ship (Naval Sea Systems Command, 2007). Thus, personnel will arrive to fill a specific job on the crew, and they will arrive fully trained.

Reconsidering IT Use Onboard

Another limitation to the most effective use of trained IT personnel is the shipboard practice of detailing the most junior personnel elsewhere on the ship. This practice was described to the research team as essential to the culture of the Navy, but it stands apart from the practice of the other services. For example, the Army does not require the most junior personnel to serve in the mess; instead, the Army contracts or enlists mess personnel. This practice was also presented as a barrier to providing more extensive training to junior sailors. In other words, why train them when the trained skills will erode before they finish their time in the mess? We did not tend to hear the opposing perspective: Why send them to mess duty after the Navy has invested in training highly specialized skills?

The LCS manning strategy addresses this disconnect between training and actual duties by assigning individuals to specific crew jobs and by defining the maintenance/crew support and watchstanding requirements for each of those jobs. Essentially, the ship's administration and supply will be performed from ashore, as will be most of the preventive maintenance, and all sailors will have to perform some tasks outside of their rating. This leaves all sailors to do their defined jobs, with all defined jobs including some non-rating work.

Movement to a Split Community with a Combination of Six-Year and Four-Year Enlistment Contracts

Another manning change in process is the change to a split community, in which some IT personnel enlist with six-year contracts while others have a four-year enlistment. This change is being made in concert with a revised training strategy for those personnel; as discussed in the next section of this chapter, those personnel will attend C school immediately following A school. The benefits of earlier C school are significant to CANES and are discussed below. The benefit of a longer enlistment is also worthwhile, as a greater percentage of personnel will have at least five years of experience. However, the proposed implementation will create a split community, which will have negative implications for the IT community itself, as it will complicate the management of the community. More importantly, it will produce a stratified community, in which some IT personnel have received additional training and thus have an advantage for promotion and retention, while other ITs do not.

Training Implications

There are several changes in training for IT personnel that have either been proposed or considered and that have implications for CANES, given that they will improve the training of IT personnel and thus increase their capability and performance on an individual level, with resulting greater effectiveness overall or—should the Navy choose to take advantage of savings gained with greater effectiveness—less IT manpower required. This section considers these changes, including the requirement for IT personnel to be IA-certified, increasing the length of IT A school, resequencing NEC training, and moving C school to earlier in the IT career.

Required Certifications

The Department of Defense has issued DoD Directive 8570.1, which documents the guidance for training, certification, and management of all government employees whose jobs include Information Assurance (IA) functions. This new directive has implications for CANES because it requires IT personnel to be IA-certified. Specifically, any personnel who work within the computing environment and have unsupervised privileged system access are required to have IA Level I training and certification. This typically includes the most junior IT personnel. More senior IT personnel who provide network environment and enhanced computing environment support, and who are especially concerned with network security, are required to have IA Level II training and certification. The most senior IT personnel, those who work within the enclave environment and on advanced network environment and advanced computing environment issues, are required to have IA Level III certification. These are civilian certifications that indicate a mastery of the information; the certifications are also meaningful in, and thus marketable to, the civilian job market.

The introduction of these required certifications has several implications for CANES. First, personnel who complete the civilian certification will bring more relevant expertise to their work within the CANES environment, as passing the civilian certification requires a demonstrated mastery of the material. Second, because these certifications are recognized and valued in the civilian job market, these personnel may have more marketable skills outside the Navy than ITs previously had, possibly reducing retention.[4]

Increasing Length of A School

The Navy has decided to extend the length of A school, both to accommodate the additional training required for the IA Level I certification and to provide some limited network training. Reflecting the additional curriculum, ITs will exit the revised A school with an IT NEC (2752). Lengthening A school with an NEC-producing curriculum is a positive move toward increasing the capability of the ITs who administer and maintain CANES. As such, it addresses some of the concerns voiced by shipboard personnel that ITs were inadequately trained prior to their first ship assignment.[5]

Resequencing NEC Training

The 2735 NEC (Journeyman Networking Core) is an important training course for IT personnel. However, as shown earlier in Figure 2.6, system-specific NECs, such as 2730 (NTCSS) and 2720 (GCCS-M System Admin) do not currently require 2735 as a prerequisite NEC. Training for the newer GCCS-M system (NEC 2710, GCCS-M 4.x System Admin) does require 2735 as prerequisite training. Prerequisite network training will be especially important as the systems are maintained and administered from the CANES environment.

Moving the network training earlier in the training pipeline could produce two positive consequences. First, training experts indicated that the system-specific training could potentially be shortened if the trainees had already completed the network training. Second, if ET 1678s are no longer required for CANES, the ITs will require some limited ET capabilities, including basic electrical safety. This 1678 material can be incorporated into the 2735 network training and should be provided to ITs prior to the completion of the system-specific training for GCCS-M or NTCSS. In summary,

---

[4] Later in the report, our assessment of moving to longer initial enlistments provides for lower six-years-of-service attrition rates as a result of enhanced training and certification.

[5] The Center for Information Dominance (CID) predicts approximately an 8 percent failure rate from the more rigorous training and specifically from the Level I certification examination. The CID personnel interviewed for this analysis agreed that a two-week remediation program could reverse the failure for approximately 50 percent of those who entered remediation. That estimate has been used in our simulation and analysis, discussed in more detail in Appendix B.

there are both efficiency and effectiveness implications to resequencing some of the NEC training for ITs.

Moving C School to the Beginning of the IT Career

The current plans for IT training include moving the IT C school to the beginning of the IT career, such that it would immediately follow A school completion, for some portion of the IT community. This would address the shipboard concerns heard during this study about the amount of training received by ITs prior to their first assignment, and it would ameliorate the contrast that was frequently made between the less trained ITs and the more capable ETs (who complete C school before the first assignment). This movement of 2735 NEC training for network ITs, and the 2379 NEC training for communications ITs, would increase considerably both the amount of overall training and the amount of hands-on training that ITs receive before their first assignment. This change would require six-year enlistment contracts for some ITs, in order to permit the time for, and recoup the investment from, the additional training.

However, the current plan is to enlist only 35 percent of the IT community with a six-year contract and to send only this portion of the community to network C school immediately following the completion of A school. The analysis supporting this Navy decision was based on the estimated cost of the implementation. In contrast, our analysis, which is based on a simulation of IT training and assignments, and which considers both the cost of the implementation and the benefits gained (trained ITs), suggests merit to enlisting all ITs with six-year contracts and sending them all to C school immediately following A school.

We provide this analysis in more detail in Appendix D, but summarize it here. The key inputs and assumptions include the length and sequencing of training and retention rates, including the difference between the retention rates of ITs with a six-year contract and ITs with a four-year contract. The key outputs of the analysis include costs (both transition and long-term) and the number or percentage of ITs assigned to billets that have received C school training. This analysis indicates that, assuming that retention drops no lower than that evidenced in the past five years, the long-term cost of sending all ITs to early C school is close to (or even less than) baseline costs. However, the benefits are tremendous: Almost all ITs assigned to a billet will have attended C school, as compared with roughly 60 percent in the base case. Further, the most junior IT personnel would be in training; the more senior IT personnel would be assigned to billets. Our analysis differs from Navy analysis of these alternatives by considering long-term, rather than just transition, costs, and by considering the benefits of the revised system.

Increased Effectiveness or Decreased Cost?

If all ITs are better trained, greater individual effectiveness is a reasonable expectation. In fact, prior research suggests that effectiveness gains of 5–15 percent are possible and reasonable.[6] There are two different kinds of gains that could be reaped from these changes. The first is that if the numbers and seniority of the assigned IT personnel remained the same, but personnel were better trained, the ITs onboard a ship associated with CANES would be more effective. Personnel at Naval Network Warfare Command (NETWARCOM) expressed support for this approach, citing concern that IT manpower requirements and associated authorizations were too low for the capabilities needed through the transition to CANES and likely thereafter.

A second possible outcome, which appeals to those who believe that the current requirements and associated authorizations are adequate to satisfy the need for ITs in the CANES environment, is to reap the effectiveness gains with reductions in the requirements and authorizations for IT personnel.

Summary

The findings of prior research suggest that there are only limited implications from CANES for the IT community. The limited effect of tremendous technological change reflects the organizational nature of the Navy, which lacks a single "czar" who can harness technological change, ship structure, manpower and manning processes, and operational practice to produce change. Because there is no single decisionmaker solely responsible for all of these aspects of the Navy, gains in efficiencies and effectiveness from the conversion to CANES will be difficult to realize fully without close collaboration and alignment of interests among the stakeholders.

Manpower reductions are feasible either through changes to legacy IT watches or through adopting a different model for calculating manpower requirements. There are also feedback loops to manpower from the changing community management and training practices. A more experienced and better-trained IT should be more productive. Not needing to supervise (or be supervised closely) frees up man-hours for productive work.

There are manning changes that may have implications for CANES. Two issues—manning to specific positions and reconsidering the use of junior personnel onboard ships—both gain insights from the manning strategy proposed for LCS. Given the optimized manning plan for LCS, the Naval Sea Systems Command (NAVSEA) has developed more-effective manning processes to eliminate wasteful use of personnel. These include finding precisely the right person for a particular position, training such individuals sufficiently en route, and sharing the common ship's duties that cannot be relegated to shore-based contractors. These guiding principles may not be appropriate to all personnel assigned to traditional Navy ships, or even to all IT personnel assigned

---

[6]   These studies are reviewed in Appendix E of Thie et al., 2009. In particular, see studies by Thomas Manacapilli and Stan Horowitz cited there.

to traditional Navy ships. Nonetheless, the training investment in IT personnel (especially if that investment increases) suggests the merit of those policies, which would likely result in more-effective IT personnel associated with CANES.

There are planned changes to IT training that have positive implications for CANES. First, the DoD requirement for personnel such as IT to be IA-certified will introduce additional training and capability to IT personnel managing and administering CANES systems. This certification is part of the reason that IT A school will be lengthened, also increasing the capability of new IT personnel. Additionally, the possible resequencing of NEC training as well as the movement of C school to the beginning of the IT career are decisions that increase the capability of IT personnel and thus result in positive outcomes for CANES. Further, the capability gains possible by ensuring that all ITs are fully trained before assignment to a unit can result in a more effective population of ITs associated with CANES, or in cost savings, if those effectiveness gains are translated to manpower reductions.

CHAPTER FOUR

Recommendations

The previous chapter presented our assessment about manpower, personnel, and training issues and our conclusions about their likely effects on CANES. This chapter summarizes our recommendations. The first recommendation is specific to manpower, personnel, and training in the CANES environment; the next six focus on ITs and thus have significant implications for CANES. The last one affects many Navy ratings and is not a new suggestion.

Recommendations

- The PEO C4I should work with NAVMAC and with organizational stakeholders (e.g., the type commanders [TYCOMs]) to either reduce watches for ITs or move to a different model for addressing manpower requirements. Ideally, the manpower model selected would permit the Navy to capitalize on technology advances, such as those resulting in improved reliability and the opportunity for virtual administration, that would otherwise suggest a reduction in manpower.
- Proceed as planned with longer A school to provide Level One IA certification to IT personnel. However, also institute a two-week remedial program for those personnel who are not initially successful with certification.
- Add critical training elements from 1678 to IT network training to facilitate the absorption of the 1678 requirement among ITs.
- Consider greater use of the LCS detailing strategy. In other words, assign IT personnel as individuals to fill specific positions, and ensure that they receive appropriate training en route.
- Enlist all IT personnel with a six-year enlistment contract and send all ITs to C school following A school, in order to dramatically increase the number of trained ITs associated with CANES.
- Explore whether early C school can reduce the length of system-specific NEC training. Additionally, if early C school is not instituted for all ITs, still consider resequencing NEC training such that network training is prerequisite for system-specific training.

- Consider whether the productivity gains from early C school should result in greater effectiveness or in manpower savings.
- Consider whether the traditional use of junior personnel onboard ships remains appropriate and effective, especially for highly trained technical personnel.

System Descriptions

Introduction

There are five C4I systems that will be among the early adopters of CANES. These systems are networks and applications of information technology used by the Navy afloat and ashore. The unique manpower and personnel requirements of each system were taken into account as we generated the forecasted changes of adapting to CANES. With IT uses and applications expanding in the Navy, these programs play a significant role, and their successful transition to the consolidated network is pivotal.

CENTRIXS-M

CENTRIXS-M (Combined Enterprise Regional Information Exchange System–Maritime) is a global network that permits information sharing through secure email, Web services, Web replication, Common Operational Picture (COP), Common Intelligence Picture (CIP), and chat function. The network provides secure tactical and operational information sharing between U.S. and coalition maritime partners.

The aim of the Web-centric network is to achieve a level of shared awareness allowing for increased speed of command. Within that goal is the integration of tactical and non-tactical LANs. CENTRIXS-M uses separate enclaves for each network connected to varying coalition partners or member nationals in order to maintain appropriate classification separation. Block II of CENTRIXS-M, however, eases communication by obviating the need to switch enclaves for each nationality with a view of up to five enclaves in a single display.

Afloat, CENTRIXS-M employs Multi-Level Thin Client (MLTC) architecture that provides the analyst with appropriate clearance and access to data stored in multiple security domains. CENTRIXS-M employs a variety of commercial off-the-shelf (COTS) software and hardware as well as government off-the-shelf (GOTS) software. As this software and hardware continuously evolves, CENTRIXS-M adapts to fielding the most up-to-date equipment.

SCI

SCI is an information-sharing network that operates at the Top Secret/SCI level. The network provides protected delivery of information through a secure network interface for cryptologic and intelligence systems. The SCI network is a tactical backbone service, using the General Services (GENSEV) Automated Digital Network Service (ADNS) to connect the user to the global information grid. SCI allows for this information sharing in ship-to-ship, ship-to-shore, and shore-to-ship interactions.

Using GOTS, COTS, and Inline Network Encryption (INE) products, SCI provides secure mechanisms for handling sensitive information. The goal of the network is to provide utilities, including file transfer, mail interfaces, interactive chat, Web services, and organizational messaging. The performance of information delivery on each of these tasks is directly related to available bandwidth, total number of active users, and the types of services being used.

This network greatly expands the capability of cryptologists and intelligence personnel to fully interact with shore- and shipboard-based systems. Analysts have increased access to situational awareness, indications and warnings, enemy force intentions, and intelligence preparation.

ISNS

The Integrated Shipboard Network System (ISNS) is a network application that enables a secure exchange of voice, video, and data between ship and shore. This exchange of voice and video merges tactical and non-tactical networks. ISNS is integrated architecturally into the Navy's end-to-end strike group. The integration capability makes ISNS able to accommodate technology refreshment and any growth. ISNS will operate in the environment of the joint community and among coalition partners as well.

The program integrates the network capabilities formerly provided by GCCS-M and NTCSS with other Navy LANs and encompasses both shipboard and embarked LANs. The consolidation of ISNS allows for standardization, more reliable and robust networking, common network management, and basic network information distribution services.

ISNS employs COTS hardware and GOTS and COTS software. It is an adaptable system able to meet changing requirements by allowing for rapid development. ISNS draws requirements from the Information Dissemination Management Requirements Document. The program is also an element of the Joint Global Information Grid.

GCCS-M

Global Command and Control System–Maritime (GCCS-M) is a system used to receive, display, correlate, and maintain geo-location data. The integration of these data with intelligence and environmental information provides a tactical picture to the analyst. Mission operations of GCCS-M include detection and display of threat information, strategy planning aid, course-of-action development, executive planning, implementation, monitoring, risk analysis, and the creation of a common tactical picture.

GCCS-M has three variants: ashore, afloat, and multilevel security. The ashore version provides C4I capability to land-based forces. It is used as the primary C4I support system and has been made available to the Joint Task Force, command centers, and NATO maritime command centers. The afloat system is the single C4I capability to sea-based forces. Its evolutionary acquisition program incorporates the functionality of many systems. Multilevel security GCCS-M enables operations in joint environments to access, retrieve, process, and disseminate all necessary information to maintain a Common Operating Picture.

This system supports the full range of imagery requirements and gives near real-time receipts and transmission of tactical imagery. GCCS-M also relies heavily on COTS products to keep up with the pace of evolution of commercial information management systems.

NTCSS

Naval Tactical Command Support System (NTCSS) is multifunctional program that is a group of software applications whose goal is to provide decision support for management of ships, submarines, aviation squadrons, and intermediate maintenance activities, both afloat and ashore. The management can include providing the unit commanding officer and crew with the ability to manage maintenance of equipment, parts inventory, finances, automated technical manuals, personnel information, medical, crews mess, ships store, and unit administrative information.

NTCSS is an integration of the SNAP (Shipboard Non-Tactical ADP [Automatic Data Processing] Program), NALCOMIS (Naval Aviation Logistics Command Management Information System), and MRMS (Maintenance Resources Management System), the three major command support programs.

The program runs on Unix and Windows New Technology (NT). Physically, it is composed primarily of COTS and Non Development Item equipment. As directed by the SPAWAR PMW 151 (Naval Combat Support Systems) program office, NTCSS has developed a series of hardware and software configurations that replace old and expensive equipment with new compliant and reliable open systems.

Training Model Description

The simulation model used for the training analysis was programmed in iThink, a system dynamics tool. The model simulates the movement of ITs as they enter the Navy and proceed through various training and assignments, until they exit the system at a retention decision point or at retirement. The model was run for over 2,500 time periods, until reaching steady state, and the steady-state patterns were used in this analysis.

This section provides an introduction to system dynamics and then explains the models used for this report.

Introduction to System Dynamics

A system dynamics model consists of stocks, flows, and auxiliaries, each of which is explained below (three models are diagrammed beginning on p. 44).

Stocks in the Model

Stocks are the primary building blocks of the model, and in this model the stocks contain ITs. Stocks appear in the model as rectangles. This model uses a variety of different types of stocks, including ovens, conveyor belts, and reservoirs. The oven stock used in the model is shown by a rectangle within a rectangle (much like an oven door with a window) and represents the recruiting mechanism. For the purpose of this model, all accessions appear once a year. Thus, the oven stock labeled "Accessions" opens its door and releases a year's worth of IT accessions, once each year.

The conveyor belt stocks have vertical stripes on the rectangles. Each of those stocks is programmed with a duration. ITs enter the conveyor belt stock from the left and, at the end of the duration, emerge from the stock at the right. For example, the conveyor belt labeled "Boot camp" has a duration of 9 weeks. Another conveyor belt with a duration of 208 weeks, or four years, represents the time ITs spend in assignments from year 9 to year 12 of their careers.

The model also includes reservoir stocks, shown as simple rectangles. In this model, reservoir stocks are used to represent the end-of-career ITs, who are counted as a pool of ITs and who attrit at a rate representing retirement.

Flows in the Model

The flows are the valves that control movement into and out of stocks. For example, the model presumes that all ITs that complete boot camp will proceed to A school, so the valve into A school is set to equal all ITs that complete boot camp. However, the model assumes that 8 percent of A school students will fail to complete the civilian certification process at the completion of A school and will thus require remediation. Thus, the valve into the remediation program is set to equal 0.08 of the number of ITs who emerge from A school. The model shows an arrow extending from the valve exiting A school to the valve entering remediation. This arrow indicates that the former is used in the formula of the latter.

Auxiliaries in the Model

Auxiliaries are represented as circles. They include information that may affect the flow or accumulation in the model. In the case of this model, they are primarily used to count ITs. For example, there is one auxiliary that counts the number of personnel in training. It adds the contents of boot camp, A school, and C school. Another auxiliary counts those ITs that are in IT training. This includes A school and C school, but not boot camp. The auxiliary does so by taking all personnel in training (from the first auxiliary) and subtracting the number of personnel in boot camp. Other auxiliaries in the model count all ITs, all ITs assigned to units, and all ITs that have attended C school and are assigned to units.

The Models Used in This Analysis

This section describes the flows through the different models used. The equations, including the retention calculations and the length of different training sessions, are provided in Appendix C.

Status Quo Model

The current training and assignment processes were modeled and labeled "status quo." This model is shown in Figure B.1, and the equations for this model are available in Appendix C. In this model, ITs access into an oven stock (Accessions) that once-annually releases ITs into boot camp. After completing boot camp, ITs progress on to A school. A portion of A school grads proceed immediately to C school, but most ITs are assigned to units after completing A school. Unit assignments in this model are identified by time and by whether the ITs have attended C school. Thus, those who have not attended C school fill their first unit assignment in the stock labeled "ITs to 4 YOS." Those who have attended C school fill assignments while in the stock named "Trained ITs to 4 YOS."

After four years of service, almost all ITs who choose to remain in the Navy but have not yet been to C school are then sent to C school. A portion of those ITs who have already been to C school are continued, consistent with retention rates, in the stock labeled "Early Trained ITs to 8 YOS." The main difference between this stock and the stock labeled "Trained ITs to 8YOS" is that the latter is shorter, representing the time that those ITs spend in C school after their fourth-year retention decision. Both of these groups encounter another retention decision at eight years of service; those who remain enter the stock "Trained 9 to 12 YOS." Meanwhile, there is another track for the very few who did not receive C school after their fourth year or who did not successfully complete C school. They proceed from "ITs 5 to 8 YOS" to "ITs 9 to 12 YOS" and eventually to "Career ITs." "Career ITs" and "Trained career ITs" contain ITs with 13 or more years of service. These ITs attrit at a rate representing eventual retirement from the community.

The figure also indicates the auxiliaries that count the numbers of different kinds of ITs. Some of the counts are done using "ghosts" of the stocks, to reduce the clutter in the model graphic.

Base Case Model

The status quo model does simulate the current practice. However, some decisions regarding changes to IT training have already been determined, such as the decision to lengthen A school in order to incorporate civilian certification in the curriculum. Because of these finalized decisions, the status quo was not a good basis for comparison. Thus, this analysis also simulated a base case representing these changes. The resulting model is shown in Figure B.2, and the equations are available in Appendix C. The base case model is very similar to the status quo model, with a few exceptions. Because the longer and more difficult A school will result in approximately 8 percent failure,[1] this model includes a two-week remediation program. The model assumes that after the remediation, half of those personnel will receive Level One certification and proceed to a unit. They are represented in the stock "Later pass to 4YOS," and the rest of their career follows the same path as ITs who initially passed A school. The ITs that do not succeed with remediation proceed to "Failed Aschool to 2." This conveyor is two years long, at which point half of these ITs have left the community. Those who remain in the community proceed to "Failed Aschool to 4," which takes them to four years of service, at which point none of them continues as an IT.

---

[1]   Estimated by CID.

Figure B.1
Status Quo Model

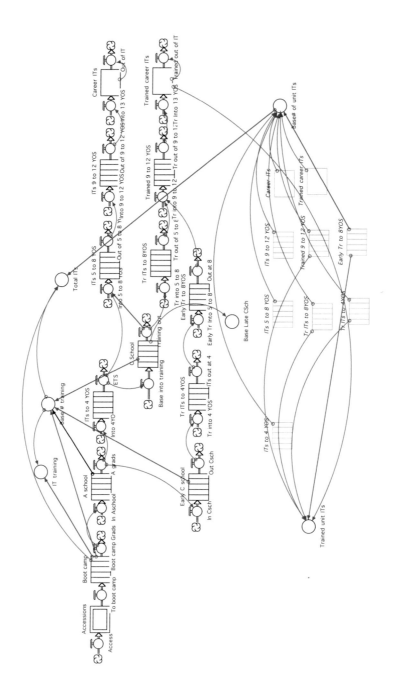

Figure B.2
Base Case Model

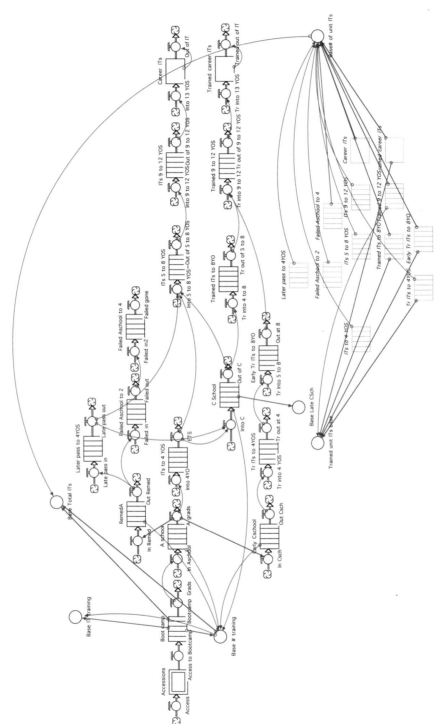

Excursions with Early C School

This analysis considered the implications of sending ITs to C school immediately following the completion of A school. This version of the model was used to evaluate multiple cases, some of which had all ITs on a six-year contract, and some of which had only the ITs attending early C school on a six-year contract. The model shown in Figure B.3 is the version used to assess all ITs on a six-year contract. The equations are available in Appendix C.

This model enters ITs as in the base case model and includes remediation for ITs that do not successfully complete A school. After A school, ITs enter one of three different conveyor stocks: network C school, communications C school, or a unit assignment. The proportion of ITs that enter each of these stocks was varied for the different analytical model runs. After successfully completing C school, ITs proceed to unit assignments. ITs who have been to C school are tracked through the model as trained ITs. Those who did not attend C school and the small proportion of ITs that do not successfully complete network C school (including Level Two certification) are tracked separately in their unit assignments. There is a retention decision after six years of service. ITs who have not attended C school previously are sent to C school if they retain past six years. The model continues ITs through their career, tracking separately those who attended C school early, those who attended C school later, and those who have not attended C school.

Retention Assumptions for This Analysis

This model uses retention at typical decision points rather than annual continuation rates. Different versions of the model had retention decisions at different times, based on whether the ITs were modeled with four years of initial service or six years of initial service. Note that all versions of the models do not have retention decisions at the same times. For example, the base case does not have six-year retention decisions. Because the community managers track continuation rates by year of service, the retention rates used for the modeling were also extrapolated into annual year-of-service continuation rates. The four-year retention probabilities were similar to recent continuation data for the IT community. Navy analysts advised that use of six-year continuation rates for the ET community as the basis for likely six-year continuation rates for the IT community was the current practice. Figure B.4 shows a five-year average for the ET continuation rates as well as the maximum and minimum rate for each of the five years. Additionally, the six-year plans were modeled with varying retention rates to assess the sensitivity of those retention assumptions. All rates were within the max and min band as shown in the figure.

Figure B.3
Early C School Excursion Model

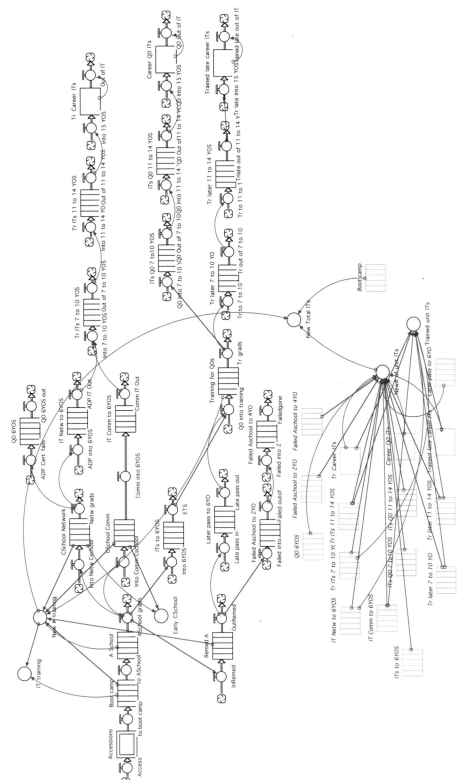

Figure B.4
ET Continuation Rates for FY2004–2008

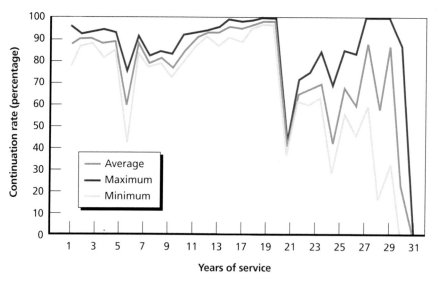

Modeling Equations

This appendix includes the equations from three of the model versions used for this analysis: the status quo model, the base case model, and the excursion with all eligible ITs immediately attending C school after A school (with moderate retention). The names of stocks and auxiliaries are highlighted with bold text. Typically, the stock equations include the definition of the stock, then the initial value, the duration, capacity, fill time, inflow and outflow, as appropriate. The auxiliary equations indicate the value or the model elements summed for the auxiliary.

Equations: Status Quo Model

Accessions(t) = Accessions(t − dt) + (Access − To_boot_camp)*dt

> INIT Accessions = 1412
>
> COOK TIME = 52
>
> CAPACITY = 1412
>
> FILL TIME = ∞
>
> INFLOWS:
>
> > Access = 1412
>
> OUTFLOWS:
>
> > To_boot_camp = CONTENTS OF OVEN AFTER COOK TIME, ZERO OTHERWISE

A_school(t) = A_school(t – dt) + (In_Aschool – A_grads)*dt

      INIT A_school = 0

      TRANSIT TIME = 11

      INFLOW LIMIT = ∞

      CAPACITY = ∞

      INFLOWS:

            In_Aschool = Boot_camp_Grads

      OUTFLOWS:

            A_grads = CONVEYOR OUTFLOW

Boot_camp(t) = Boot_camp(t – dt) + (To_boot_camp – Boot_camp_Grads)*dt

      INIT Boot_camp = 0

      TRANSIT TIME = 9

      INFLOW LIMIT = ∞

      CAPACITY = ∞

      INFLOWS:

            To_boot_camp = CONTENTS OF OVEN AFTER COOK TIME, ZERO OTHERWISE

      OUTFLOWS:

            Boot_camp_Grads = CONVEYOR OUTFLOW

Career_ITs(t) = Career_ITs(t – dt) + (Into_13_YOS – Out_of_IT)*dt

      INIT Career_ITs = 0

      INFLOWS:

            Into_13_YOS = 0.75*Out_of_9_to_12_YOS

      OUTFLOWS:

            Out_of_IT = 0.0035*Career_ITs

C_School(t) = C_School(t – dt) + (Base_into_training – Training_out)*dt

 INIT C_School = 0

   TRANSIT TIME = 14

   INFLOW LIMIT = ∞

   CAPACITY = ∞

 INFLOWS:

  Base_into_training = IF ETS=0 THEN 0 ELSE (0.48*ETS)

 OUTFLOWS:

  Training_out = CONVEYOR OUTFLOW

Early_C_school(t) = Early_C_school(t – dt) + (In_Csch – Out_Csch)*dt

 INIT Early_C_school = 0

 TRANSIT TIME = 14

 INFLOW LIMIT = ∞

 CAPACITY = ∞

 INFLOWS:

  In_Csch = 0.19*A_grads

 OUTFLOWS:

  Out_Csch = CONVEYOR OUTFLOW

Early_Tr_to_8YOS(t) = Early_Tr_to_8YOS(t – dt) + (Early_Tr_Into_5_to_8 – Out_at_8)*dt

 INIT Early_Tr_to_8YOS = 0

 TRANSIT TIME = 208

 INFLOW LIMIT = ∞

 CAPACITY = ∞

 INFLOWS:

  Early_Tr_Into_5_to_8 = 0.5*ITs_out_at_4

 OUTFLOWS:

  Out_at_8 = CONVEYOR OUTFLOW

ITs_5_to_8_YOS(t) = ITs_5_to_8_YOS(t – dt) +
(Into_5_to_8_YOS – Out_of_5_to_8_YOS)*dt

   INIT ITs_5_to_8_YOS = 0

   TRANSIT TIME = 110

   INFLOW LIMIT = ∞

   CAPACITY = ∞

   INFLOWS:

      Into_5_to_8_YOS = (0.02*ETS) + (0.03*Training_out)

   OUTFLOWS:

      Out_of_5_to_8_YOS = CONVEYOR OUTFLOW

ITs_9_to_12_YOS(t) = ITs_9_to_12_YOS(t – dt) +
(Into_9_to_12_YOS – Out_of_9_to_12_YOS)*dt

   INIT ITs_9_to_12_YOS = 0

   TRANSIT TIME = 208

   INFLOW LIMIT = ∞

   CAPACITY = ∞

   INFLOWS:

      Into_9_to_12_YOS = (0.45*Out_of_5_to_8_YOS)

   OUTFLOWS:

      Out_of_9_to_12_YOS = CONVEYOR OUTFLOW

ITs_to_4_YOS(t) = ITs_to_4_YOS(t – dt) + (Into_4YOS – ETS)*dt

   INIT ITs_to_4_YOS = 0

   TRANSIT TIME = 188

   INFLOW LIMIT = ∞

   CAPACITY = ∞

   INFLOWS:

      Into_4YO = 0.81*A_grads

   OUTFLOWS:

      ETS = CONVEYOR OUTFLOW

Trained_9_to_12_YOS(t) = Trained_9_to_12_YOS(t – dt) + (Tr_into_9_to_12 – Tr_out_of_9_to_12)*dt

>   INIT Trained_9_to_12_YOS = 0

>   TRANSIT TIME = 194

>   INFLOW LIMIT = ∞

>   CAPACITY = ∞

>   INFLOWS:

>>   Tr_into_9_to_12 = (0.45*Tr_out_of_5_to_8) + (0.45*Out_at_8)

>   OUTFLOWS:

>>   Tr_out_of_9_to_12 = CONVEYOR OUTFLOW

Trained_career_ITs(t) = Trained_career_ITs(t – dt) + (Tr_Into_13_YOS – Trained_out_of_IT)*dt

>   INIT Trained_career_ITs = 0

>   INFLOWS:

>>   Tr_Into_13_YOS = 0.75*Tr_out_of_9_to_12

>   OUTFLOWS:

>>   Trained_out_of_IT = 0.0035*Trained_career_ITs

Tr_ITs_to_4YOS(t) = Tr_ITs_to_4YOS(t – dt) + (Tr_into_4_YOS – ITs_out_at_4)*dt

>   INIT Tr_ITs_to_4YOS = 0

>   TRANSIT TIME = 174

>   INFLOW LIMIT = ∞

>   CAPACITY = ∞

>   INFLOWS:

>>   Tr_into_4_YOS = Out_Csch

>   OUTFLOWS:

>>   ITs_out_at_4 = CONVEYOR OUTFLOW

Tr_ITs_to_8YOS(t) = Tr_ITs_to_8YOS(t – dt) + (Tr_into_5_to_8 – Tr_out_of_5_to_8)*dt

> INIT Tr_ITs_to_8YOS = 0

> TRANSIT TIME = 194

> INFLOW LIMIT = ∞

> CAPACITY = ∞

> INFLOWS:

>> Tr_into_5_to_8 = 0.97*Training_out

> OUTFLOWS:

>> Tr_out_of_5_to_8 = CONVEYOR OUTFLOW

Base#_of_unit_ITs = ITs_to_4_YOS + ITs_5_to_8_YOS + ITs_9_to_12_YOS + Career_ITs + Trained_9_to_12_YOS + Trained_career_ITs + Tr_ITs_to_8YOS + Tr_ITs_to_4YOS + Early_Tr_to_8YOS

Base_#_training = Boot_camp + A_school + C_School + Early_C_school

Base_Late_CSch = C_School

IT_training = Base_#_training – Boot_camp

Total_ITs = Base#_of_unit_ITs + Base_#_training – Boot_camp

Trained_unit_ITs = Tr_ITs_to_8YOS + Trained_9_to_12_YOS + Trained_career_ITs + Tr_ITs_to_4YOS + Early_Tr_to_8YOS

Equations: Base Case Model

Accessions(t) = Accessions(t – dt) + (Access – Access_to_Bootcamp)*dt

        INIT Accessions = 1481

        COOK TIME = 52

        CAPACITY = 1481

        FILL TIME = ∞

        INFLOWS:

                Access = 1481

        OUTFLOWS:

                Access_to_Bootcamp = CONTENTS OF OVEN AFTER COOK TIME, ZERO OTHERWISE

A_school(t) = A_school(t – dt) + (In_Aschool – A_grads)*dt

        INIT A_school = 0

        TRANSIT TIME = 19

        INFLOW LIMIT = ∞

        CAPACITY = ∞

        INFLOWS:

                In_Aschool = Bootcamp_Grads

        OUTFLOWS:

                A_grads = CONVEYOR OUTFLOW

Boot_camp(t) = Boot_camp(t – dt) + (Access_to_Bootcamp – Bootcamp_Grads)*dt

      INIT Boot_camp = 0

      TRANSIT TIME = 9

      INFLOW LIMIT = ∞

      CAPACITY = ∞

      INFLOWS:

            Access_to_Bootcamp = CONTENTS OF OVEN AFTER COOK TIME, ZERO OTHERWISE

      OUTFLOWS:

            Bootcamp_Grads = CONVEYOR OUTFLOW

Career_ITs(t) = Career_ITs(t – dt) + (Into_13_YOS – Out_of_IT)*dt

      INIT Career_ITs = 0

      INFLOWS:

            Into_13_YOS = 0.75*Out_of_9_to_12_YOS

      OUTFLOWS:

            Out_of_IT = 0.0035*Career_ITs

C_School(t) = C_School(t – dt) + (Into_C – Out_of_C)*dt

      INIT C_School = 0

      TRANSIT TIME = 16

      INFLOW LIMIT = ∞

      CAPACITY = ∞

      INFLOWS:

            Into_C = (0.48*ETS) + (0.48*Late_pass_out)

      OUTFLOWS:

            Out_of_C = CONVEYOR OUTFLOW

Early_Cschool(t) = Early_Cschool(t – dt) + (In_Csch – Out_Csch)*dt

      INIT Early_Cschool = 0

      TRANSIT TIME = 16

      INFLOW LIMIT = ∞

      CAPACITY = ∞

      INFLOWS:

            In_Csch = 0.19*A_grads

      OUTFLOWS:

            Out_Csch = CONVEYOR OUTFLOW

Early_Tr_ITs_to_8YO(t) = Early_Tr_ITs_to_8YO(t – dt) + (Tr_Into_5_to_8 – Out_at_8)*dt

      INIT Early_Tr_ITs_to_8YO = 0

      TRANSIT TIME = 208

      INFLOW LIMIT = ∞

      CAPACITY = ∞

      INFLOWS:

            Tr_Into_5_to_8 = 0.5*Tr_out_at_4

      OUTFLOWS:

            Out_at_8 = CONVEYOR OUTFLOW

Failed_Aschool_to_2(t) = Failed_Aschool_to_2(t – dt) + (Failed_in – Failed_out)*dt

      INIT Failed_Aschool_to_2 = 0

      TRANSIT TIME = 74

      INFLOW LIMIT = ∞

      CAPACITY = ∞

      INFLOWS:

            Failed_in = 0.5*Out_Remed

      OUTFLOWS:

            Failed_out = CONVEYOR OUTFLOW

Failed_Aschool_to_4(t) = Failed_Aschool_to_4(t – dt) + (Failed_in2 – Failed_gone)*dt

>   INIT Failed_Aschool_to_4 = 0

>   TRANSIT TIME = 104

>   INFLOW LIMIT = ∞

>   CAPACITY = ∞

>   INFLOWS:

>>   Failed_in2 = 0.5*Failed_out

>   OUTFLOWS:

>>   Failed_gone = CONVEYOR OUTFLOW

ITs_5_to_8_YOS(t) = ITs_5_to_8_YOS(t – dt) + (Into_5_to_8_YOS – Out_of_5_to_8_YOS)*dt

>   INIT ITs_5_to_8_YOS = 0

>   TRANSIT TIME = 208

>   INFLOW LIMIT = ∞

>   CAPACITY = ∞

>   INFLOWS:

>>   Into_5_to_8_YOS = (0.02*ETS) + (0.02*Late_pass_out) + (0.03*Out_of_C)

>   OUTFLOWS:

>>   Out_of_5_to_8_YOS = CONVEYOR OUTFLOW

ITs_9_to_12_YOS(t) = ITs_9_to_12_YOS(t – dt) + (Into_9_to_12_YOS – Out_of_9_to_12_YOS)*dt

> INIT ITs_9_to_12_YOS = 0
>
> TRANSIT TIME = 208
>
> INFLOW LIMIT = ∞
>
> CAPACITY = ∞
>
> INFLOWS:
>
>> Into_9_to_12_YOS = (0.45*Out_of_5_to_8_YOS)
>
> OUTFLOWS:
>
>> Out_of_9_to_12_YOS = CONVEYOR OUTFLOW

ITs_to_4_YOS(t) = ITs_to_4_YOS(t – dt) + (Into_4YO – ETS)*dt

> INIT ITs_to_4_YOS = 0
>
> TRANSIT TIME = 180
>
> INFLOW LIMIT = ∞
>
> CAPACITY = ∞
>
> INFLOWS:
>
>> Into_4YO = 0.73*A_grads
>
> OUTFLOWS:
>
>> ETS = CONVEYOR OUTFLOW

Later_pass_to_4YOS(t) = Later_pass_to_4YOS(t – dt) + (Late_pass_in – Late_pass_out)*dt

> INIT Later_pass_to_4YOS = 0
>
> TRANSIT TIME = 178
>
> INFLOW LIMIT = ∞
>
> CAPACITY = ∞
>
> INFLOWS:
>
>> Late_pass_in = 0.5*Out_Remed
>
> OUTFLOWS:
>
>> Late_pass_out = CONVEYOR OUTFLOW

RemedA(t) = RemedA(t – dt) + (In_Remed – Out_Remed)*dt

      INIT RemedA = 0

      TRANSIT TIME = 2

      INFLOW LIMIT = ∞

      CAPACITY = ∞

      INFLOWS:

            In_Remed = 0.08*A_grads

      OUTFLOWS:

            Out_Remed = CONVEYOR OUTFLOW

Trained_9_to_12_YOS(t) = Trained_9_to_12_YOS(t – dt) + (Tr_into_9_to_12 – Tr_out_of_9_to_12_YOS)*dt

      INIT Trained_9_to_12_YOS = 0

      TRANSIT TIME = 208

      INFLOW LIMIT = ∞

      CAPACITY = ∞

      INFLOWS:

            Tr_into_9_to_12 = (0.45*Tr_out_of_5_to_8) + (0.45*Out_at_8)

      OUTFLOWS:

            Tr_out_of_9_to_12_YOS = CONVEYOR OUTFLOW

Trained_career_ITs(t) = Trained_career_ITs(t – dt) + (Tr_Into_13_YOS – Trained_out_of_IT)*dt

      INIT Trained_career_ITs = 0

      INFLOWS:

            Tr_Into_13_YOS = 0.75*Tr_out_of_9_to_12_YOS

      OUTFLOWS:

            Trained_out_of_IT = 0.0035*Trained_career_ITs

Trained_ITs_to_8YO(t) = Trained_ITs_to_8YO(t – dt) + (Tr_in_5_to_8 – Tr_out_of_5_to_8)*dt

>   INIT Trained_ITs_to_8YOS = 0

>   TRANSIT TIME = 192

>   INFLOW LIMIT = ∞

>   CAPACITY = ∞

>   INFLOWS:

>>   Tr_in_5_to_8 = 0.97*Out_of_C

>   OUTFLOWS:

>>   Tr_out_of_5_to_8 = CONVEYOR OUTFLOW

Tr_ITs_to_4YOS(t) = Tr_ITs_to_4YOS(t – dt) + (Tr_into_4_YOS – Tr_out_at_4)*dt

>   INIT Tr_ITs_to_4YOS = 0

>>   TRANSIT TIME = 172

>>   INFLOW LIMIT = ∞

>>   CAPACITY = ∞

>   INFLOWS:

>>   Tr_into_4_YOS = Out_Csch

>   OUTFLOWS:

>>   Tr_out_at_4 = CONVEYOR OUTFLOW

Base#_of_unit_ITs = ITs_to_4_YOS + ITs_5_to_8_YOS + ITs_9_to_12_YOS + Career_ITs + Trained_9_to_12_YOS + Trained_career_ITs + Trained_ITs_to_8YO + Tr_ITs_to_4YOS + Early_Tr_ITs_to_8YO + Failed_Aschool_to_2 + Failed_Aschool_to_4 + Later_pass_to_4YOS

Base_#_training = Boot_camp + A_school + C_School + Early_Cschool + RemedA

Base_IT_training = Base_#_training – Boot_camp

Base_Late_CSch = C_School

Base_Total_ITs = Base#_of_unit_ITs + Base_#_training − Boot_camp

Trained_unit_ITs_base = Trained_ITs_to_8YO + Trained_9_to_12_YOS + Trained_career_ITs + Tr_ITs_to_4YOS + Early_Tr_ITs_to_8YO

Equations: Early C School Excursion Model

Accessions(t) = Accessions(t − dt) + (Access − to_boot_camp)*dt

      INIT Accessions = 1160

      COOK TIME = 52

      CAPACITY = 1160

      FILL TIME = DT

      INFLOWS:

            Access = 1160

      OUTFLOWS:

            to_boot_camp = CONTENTS OF OVEN AFTER COOK TIME, ZERO OTHERWISE

A_School(t) = A_School(t − dt) + (to_ASchool − ASchool_grads)*dt

      INIT A_School = 0

      TRANSIT TIME = 19

      INFLOW LIMIT = ∞

      CAPACITY = ∞

      INFLOWS:

            to_ASchool = CONVEYOR OUTFLOW

      OUTFLOWS:

            ASchool_grads = CONVEYOR OUTFLOW

Boot_camp(t) = Boot_camp(t – dt) + (to_boot_camp – to_ASchool)*dt

> INIT Boot_camp = 0
>
> TRANSIT TIME = 9
>
> INFLOW LIMIT = ∞
>
> CAPACITY = ∞
>
> INFLOWS:
>
>> to_boot_camp = CONTENTS OF OVEN AFTER COOK TIME, ZERO OTHERWISE
>
> OUTFLOWS:
>
>> to_ASchool = CONVEYOR OUTFLOW

Career_Q0_ITs(t) = Career_Q0_ITs(t – dt) + (Q0_Into_15_YOS – Q0_Out_of_IT)*dt

> INIT Career_Q0_ITs = 0
>
> INFLOWS:
>
>> Q0_Into_15_YOS = 0.85*Q0_Out_of11_to_14_YO
>
> OUTFLOWS:
>
>> Q0_Out_of_IT = 0.0035*Career_Q0_ITs

CSchool_Comm(t) = CSchool_Comm(t – dt) + (Into_Comm_Cschool – Comm_into_6YOS)*dt

> INIT CSchool_Comm = 0
>
> TRANSIT TIME = 15
>
> INFLOW LIMIT = ∞
>
> CAPACITY = ∞
>
> INFLOWS:
>
>> Into_Comm_Cschool = 0.37*Aschool_grads
>
> OUTFLOWS:
>
>> Comm_into_6YOS = CONVEYOR OUTFLOW

CSchool_Network(t) = CSchool_Network(t – dt) + (Into_Netw_Cschool – Netw_grads)*dt

> INIT CSchool_Network = 0
>
> TRANSIT TIME = 16
>
> INFLOW LIMIT = ∞
>
> CAPACITY = ∞
>
> INFLOWS:
>
>> Into_Netw_Cschool = 0.55*Aschool_grads
>
> OUTFLOWS:
>
>> Netw_grads = CONVEYOR OUTFLOW

Failed_Aschool_to_2YO(t) = Failed_Aschool_to_2YO(t – dt) + (Failed_into – Failed_outof)*dt

> INIT Failed_Aschool_to_2YO = 0
>
> TRANSIT TIME = 74
>
> INFLOW LIMIT = ∞
>
> CAPACITY = ∞
>
> INFLOWS:
>
>> Failed_into = 0.5*OutRemed
>
> OUTFLOWS:
>
>> Failed_outof = CONVEYOR OUTFLOW

Failed_Aschool_to_4YO(t) = Failed_Aschool_to_4YO(t – dt) + (Failed_into_2 – Failedgone)*dt

> INIT Failed_Aschool_to_4YO = 0
>
> TRANSIT TIME = 104
>
> INFLOW LIMIT = ∞
>
> CAPACITY = ∞
>
> INFLOWS:
>
>> Failed_into_2 = 0.5*Failed_outof
>
> OUTFLOWS:
>
>> Failedgone = CONVEYOR OUTFLOW

ITs_Q0_11_to_14_YOS(t) = ITs_Q0_11_to_14_YOS(t – dt) + (Q0_Into_11_to_14_YOS – Q0_Out_of11_to_14_YO)*dt

INIT ITs_Q0_11_to_14_YOS = 0

TRANSIT TIME = 208

INFLOW LIMIT = ∞

CAPACITY = ∞

INFLOWS:

Q0_Into_11_to_14_YOS = (0.47*Q0_Out_of_7_to_10_YOS)

OUTFLOWS:

Q0_Out_of11_to_14_YO = CONVEYOR OUTFLOW

ITs_Q0_7_to10_YOS(t) = ITs_Q0_7_to10_YOS(t – dt) + (Q0_Into_7_to_10_YOS – Q0_Out_of_7_to_10_YOS)*dt

INIT ITs_Q0_7_to10_YOS = 0

TRANSIT TIME = 192

INFLOW LIMIT = ∞

CAPACITY = ∞

INFLOWS:

Q0_Into_7_to_10_YOS = 0.03*Tr_grads

OUTFLOWS:

Q0_Out_of_7_to_10_YOS = CONVEYOR OUTFLOW

ITs_to_6YOS(t) = ITs_to_6YOS(t – dt) + (Into_6YOS – ETS)*dt

INIT ITs_to_6YOS = 0

TRANSIT TIME = 180

INFLOW LIMIT = ∞

CAPACITY = ∞

INFLOWS:

Into_6YOS = 0*Aschool_grads

OUTFLOWS:

ETS = CONVEYOR OUTFLOW

IT_Comm_to_6YOS(t) = IT_Comm_to_6YOS(t – dt) + (Comm_into_6YOS – Comm_IT_Out)*dt

> INIT IT_Comm_to_6YOS = 0
>
> TRANSIT TIME = 269
>
> INFLOW LIMIT = ∞
>
> CAPACITY = ∞
>
> INFLOWS:
>
> > Comm_into_6YOS = CONVEYOR OUTFLOW
>
> OUTFLOWS:
>
> > Comm_IT_Out = CONVEYOR OUTFLOW

IT_Netw_to_6YOS(t) = IT_Netw_to_6YOS(t – dt) + (ADP_into_6YOS – ADP_IT_Out)*dt

> INIT IT_Netw_to_6YOS = 0
>
> TRANSIT TIME = 268
>
> INFLOW LIMIT = ∞
>
> CAPACITY = ∞
>
> INFLOWS:
>
> > ADP_into_6YOS = 0.975*Netw_grads
>
> OUTFLOWS:
>
> > ADP_IT_Out = CONVEYOR OUTFLOW

Later_pass_to_6YO(t) = Later_pass_to_6YO(t – dt) + (Late_pass_in – Late_pass_out)*dt

> INIT Later_pass_to_6YO = 0
>
> TRANSIT TIME = 282
>
> INFLOW LIMIT = ∞
>
> CAPACITY = ∞
>
> INFLOWS:
>
> > Late_pass_in = 0.5*OutRemed
>
> OUTFLOWS:
>
> > Late_pass_out = CONVEYOR OUTFLOW

QO_6YOS(t) = QO_6YOS(t – dt) + (ADP_Cert_fails – QO_6YOS_out)*dt

      INIT QO_6YOS = 0

      TRANSIT TIME = 268

      INFLOW LIMIT = ∞

      CAPACITY = ∞

      INFLOWS:

            ADP_Cert_fails = 0.025*Netw_grads

      OUTFLOWS:

            QO_6YOS_out = CONVEYOR OUTFLOW

Remed_A(t) = Remed_A(t – dt) + (InRemed – OutRemed)*dt

      INIT Remed_A = 0

      TRANSIT TIME = 2

      INFLOW LIMIT = ∞

      CAPACITY = ∞

      INFLOWS:

            InRemed = 0.08*ASchool_grads

      OUTFLOWS:

            OutRemed = CONVEYOR OUTFLOW

Trained_late_career_ITs(t) = Trained_late_career_ITs(t – dt) + (Tr_late_Into_15_YOS – Trained_late_out_of_IT)*dt

      INIT Trained_late_career_ITs = 0

      INFLOWS:

            Tr_late_Into_15_YOS = 0.85*Trlate_out_of_11_to_14_YOS

      OUTFLOWS:

            Trained_late_out_of_IT = 0.0035*Trained_late_career_ITs

Training_for_Q0s(t) = Training_for_Q0s(t – dt) + (Q0_into_training – Tr_grads)*dt

INIT Training_for_Q0s = 0

TRANSIT TIME = 16

INFLOW LIMIT = ∞

CAPACITY = ∞

INFLOWS:

Q0_into_training = (0.50*ETS) + (0.5*Late_pass_out)

OUTFLOWS:

Tr_grads = CONVEYOR OUTFLOW

Tr_Career_ITs(t) = Tr_Career_ITs(t – dt) + (Into_15_YOS – Out_of_IT)*dt

INIT Tr_Career_ITs = 0

INFLOWS:

Into_15_YOS = 0.85*Out_of_11_to_14_YOS

OUTFLOWS:

Out_of_IT = 0.0035*Tr_Career_ITs

Tr_ITs_11_to_14_YOS(t) = Tr_ITs_11_to_14_YOS(t – dt) +
(Into_11_to_14_YOS – Out_of_11_to_14_YOS)*dt

INIT Tr_ITs_11_to_14_YOS = 0

TRANSIT TIME = 208

INFLOW LIMIT = ∞

CAPACITY = ∞

INFLOWS:

Into_11_to_14_YOS = 0.47*Out_of_7_to_10_YOS

OUTFLOWS:

Out_of_11_to_14_YOS = CONVEYOR OUTFLOW

Tr_ITs_7_to_10_YOS(t) = Tr_ITs_7_to_10_YOS(t – dt) + (Into_7_to_10_YOS – Out_of_7_to_10_YOS)*dt

INIT Tr_ITs_7_to_10_YOS = 0

TRANSIT TIME = 208

INFLOW LIMIT = ∞

CAPACITY = ∞

INFLOWS:

Into_7_to_10_YOS = 0.5*(ADP_IT_Out + Comm_IT_Out)

OUTFLOWS:

Out_of_7_to_10_YOS = CONVEYOR OUTFLOW

Tr_later_11_to_14_YOS(t) = Tr_later_11_to_14_YOS(t – dt) + (Tr_to_11_to_14 – Trlate_out_of_11_to_14_YOS)*dt

INIT Tr_later_11_to_14_YOS = 0

TRANSIT TIME = 208

INFLOW LIMIT = ∞

CAPACITY = ∞

INFLOWS:

Tr_to_11_to_14 = 0.47*Tr_out_of_7_to_10

OUTFLOWS:

Trlate_out_of_11_to_14_YOS = CONVEYOR OUTFLOW

Tr_later_7_to_10_YO(t) = Tr_later_7_to_10_YO(t – dt) + (Tr_to_7_to_10 – Tr_out_of_7_to_10)*dt

INIT Tr_later_7_to_10_YO = 0

TRANSIT TIME = 192

INFLOW LIMIT = ∞

CAPACITY = ∞

INFLOWS:

Tr_to_7_to_10 = 0.97*Tr_grads

OUTFLOWS:

Tr_out_of_7_to_10 = CONVEYOR OUTFLOW

Early_CSchool = CSchool_Network + CSchool_Comm

IT_training = New_#_training – Boot_camp

New#_of_unit_ITs = Career_Q0_ITs + ITs_Q0_7_to10_YOS + ITs_Q0_11_to_14_YOS + IT_Netw_to_6YOS + IT_Comm_to_6YOS + ITs_to_6YOS + Trained_late_career_ITs + Tr_Career_ITs + Tr_ITs_7_to_10_YOS + Tr_ITs_11_to_14_YOS + Tr_later_7_to_10_YO + Tr_later_11_to_14_YOS + QO_6YOS + Failed_Aschool_to_2YO + Failed_Aschool_to_4YO + Later_pass_to_6YO

New_#_training = Boot_camp + A_School + CSchool_Network + CSchool_Comm + Training_for_Q0s + Remed_A

New_Total_ITs = New#_of_unit_ITs + New_#_training – Boot_camp

Trained_unit_ITs = IT_Netw_to_6YOS + IT_Comm_to_6YOS + Trained_late_career_ITs + Tr_Career_ITs + Tr_ITs_7_to_10_YOS + Tr_ITs_11_to_14_YOS + Tr_later_7_to_10_YO + Tr_later_11_to_14_YOS

Benefits and Costs

Initial Analysis

Figure D.1 provides a summary of our analysis of training options for the IT community. This analysis included five cases. The Status Quo case represents the current management of ITs. Because the Navy has committed to changing the current training process, to lengthen the A school to 19 weeks (from 11), we have a Base Case reflecting these planned changes. In the Status Quo and the Base Case, all ITs enter on a four-years-of-service (4YOS) contract. In Case 1, which reflects current plans for ITs, the IT community is split between those entering with 4YOS and those entering with 6YOS contracts. Thirty-five percent of the community enters with a 6YOS contract and attends C school immediately after A school. Case 2 increases the portion of the ITs that enter with 6YOS contracts to 60 percent, and 35 percent of the community enters with a 6YOS and attends network C school immediately following A school. In this case, another 25 percent enters with a 6YOS contract and attends communications C school immediately after A school. In the final case, all ITs are 6YOS, and they are split such that virtually all (92 percent) of them attend either the network or the communications C school.

Figure D.1 shows that a significant increase in initially C school–trained ITs can be gained for a slight increase in training load and with fewer accessions. The benefit is obtained as a result of increasing the initial commitment to 6YOS and then providing formal schooling to virtually all of the new entrants. This contrasts with a four-year commitment and some initial C school (more C school later) in the status quo and base case and a mix of four- and six-year commitments and various proportions of C school–trained entrants in the other options. This is a steady-state analysis that shows the outcomes of the system after transition effects have worked through. There is an important distinction that does not show in the data. For the status quo and the base case, most of C school training takes place after the first assignment. For the other cases, more of the C school training is done at initial entry (before the first assignment), and for the last case it is virtually all accomplished at initial entry. Not only are more trained overall, but they are trained sooner as one moves from left to right in Figure D.1.

Figure D.1
Analysis of IT Training Options

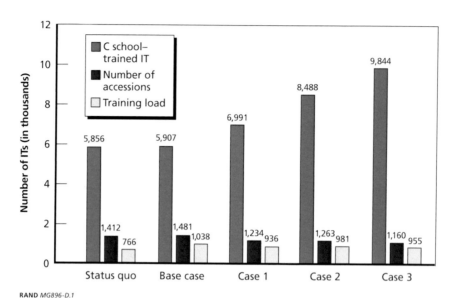

RAND *MG896-D.1*

The results in Figure D.1 are heavily influenced by retention rates for 4YOS and 6YOS personnel. Our analysis uses retention behaviors at certain decision points in a career timeline rather than year-of-service continuation rates. Our four-year retention behavior is consistent with 4YOS IT continuation rates. We tested our six-year retention behaviors against continuation rates for 6YOS personnel in the ET community as follows. For the last five years of ET continuation data, we calculated the maximum, minimum, and average rate for each year of service. Continuation rates corresponding to our periodic retention rates are within the min and max bounds of the historical data. Nevertheless, we redid our analysis using successively lower retention rates, and this analysis is discussed later in this appendix.

Further Analysis

We were asked to develop the data and analysis for a graphic that showed a scale similar to the one in Figure D.1 but defined in terms of dollars invested versus potential for return. For example, "a $XX level of investment now in training gets you $YY in return long term, and the potential for reallocation/alignment of ZZ billets." This section develops the analysis for that graphic.

The analysis requires assumptions and data about both steady-state costs and benefits (long term) and transition costs (short term). We analyzed the base case and two others from Figure D.1.

We estimated costs on a rough order of magnitude basis.[1] The costs are those for changes in training and those for changes in manpower (either number of billets or the cost of the billets). We estimated variable costs of the different cases. If there are one-time (start-up) fixed costs, they are not included, but they could be. Table D.1 summarizes those costs.

Initial Costs for Each Case

Figure D.2 shows the initial annual estimated training costs for each case over a 15-year horizon. The base case establishes the costs with which the excursions are compared. The annual cost of approximately $108 million has two significant parts: the cost of boot camp ($24 million) and the training and personnel costs of initial A school and a later C school ($84 million).[2] Case 1, in which 35 percent of new entrants are 6YOS and immediately go to C school, has reduced steady-state costs ($94 million) for the following reasons. First, the additional years of service reduce the need for new accessions and for number of people to be trained. Moreover, some of the personnel in the Individuals Account are now junior personnel, compared with more senior personnel who attend C school at the fourth or fifth year of service; their cost is less. However, there is an increase in cost above the steady state for the first four years due to the additional C school training to migrate to the new steady state.[3]

Case 3, in which all new entrants are 6YOS and virtually all immediately go to C school, has reduced steady-state costs ($87 million) for the following reasons. First, even fewer accessions exist than with Case 1. All of the entrants are initially C school–trained, so the cost per person in the Individuals Account decreases. This is a large part of the cost difference and is discussed further below. There are larger transition costs to move to the new steady state.

Overall, aggregating costs on a present value basis for 15 years for the three cases, Case 3 is the least expensive ($1.11 billion) and represents steady-state savings over Case 1 ($1.14 billion) and the Base Case ($1.29 billion). There is an upfront investment in training and transition that yields a long-term benefit.

---

[1]   We estimated these costs from other training studies or took them from cost databases maintained by the service or DoD.

[2]   Note that we are not costing the entire cost of the IT community but only the cost of that community when it is in a formal training program and thus carried in the Individuals Account for program and budget purposes. Moreover, we use this account as a proxy for the number of billets that are not available to operational forces even if they are not formally in the Individuals Account. The size and composition of the IT community does not change, only the accounting for it. We relax this condition later in the analysis.

[3]   Because the transition increases the number of people in training (Individuals Account) at the expense of the operating account, we consider this to be the cost of hiring contractors in the shore establishment to make up the gap.

Table D.1
Costing Considerations

Variable	Definition	Value
USN IA SS	Number in Individuals Account due to boot camp	Case-dependent
IT IA SS	Number in Individuals Account due to A or C school	Case-dependent
TRANSITION IT IA	Number in Individuals Account due to increase in C school during transition	Case-dependent
ACCESSIONS	Number of new entrants	Case-dependent
COST TNG WEEK, A OR C	Excludes student personnel costs	$1,000
COST WEEK BOOT CAMP	Includes new entrant personnel costs	$1,800
MPN COST EARLY	Cost of an E-2	$40,000
MPN COST LATE	Cost of an E-5	$100,000
ANNUAL RECRUITS PER RECRUITER	High-quality recruits per recruiter per year	18
USN IA SS COST	Annual cost for boot camp for new entrants	Outcome
IT IA SS COST	Annual cost for A and C school training; includes personnel and training cost	Outcome
TRANSITION IT IA COST	Annual cost for A and C school training during transition; includes personnel and training cost	Outcome
ACCESSION COST DIFFERENCE	For changed accessions	Outcome
TOTAL (15 YR HORIZON)	Net present value	Outcome

NOTE: Training week cost is estimated from discussions with training personnel and a prior study of other Navy specialties (see studies and sources cited in Thie et al., 2009). Estimated boot camp cost is derived from that estimate, presuming the need for less technology and equipment in training. The costs of E-2 and E-5 personnel are from DoD programming guidance. The annual number of high-quality recruits per recruiter is an estimate based on recruiter productivity studies (see, for example, Dertouzos and Garber, 2006).

Figure D.3 shows the other benefit from making the investment. Virtually all ITs have C school training and virtually all have it as part of their initial training. The investment yields lower steady-state costs and better-trained personnel.

Figure D.4 summarizes aggregate cost for 5-, 10-, and 15-year periods on a present value basis. Both Case 1 and Case 3 are similar in their savings for all periods. However, Case 3 has the significant advantages shown in Figure D.3: Virtually all ITs have initial C school training before their first assignment.

Reaping the Dollar Savings

The increase in trained personnel occurs given decisions to implement the change to a 6YOS community and initial C school training. However, some of the dollar savings represent "analytical" savings without a parallel in the real world of program and budget. In particular, the savings attributed to a less expensive billet or space in the

Figure D.2
Initial Costs for Each Case

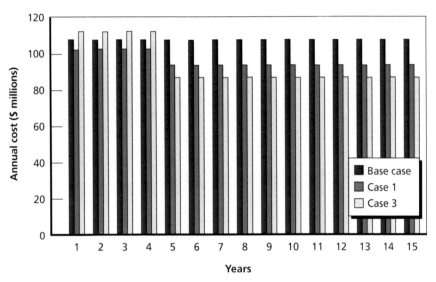

Figure D.3
C School–Trained Personnel

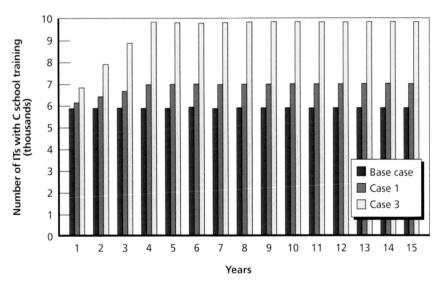

Figure D.4
Aggregate Cost Over Time

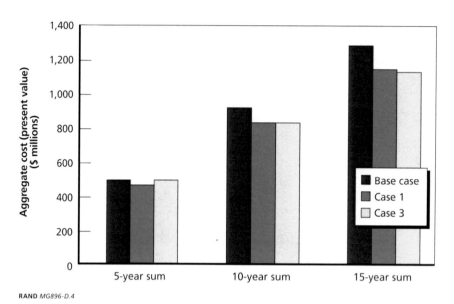

Individuals Account does not change the size or composition of the IT community. In reality, the cost of an operational billet increases (has more E-5s and E-6s) and the cost of an individual account billet decreases (fewer E-5s and E-6s), which accounts for a large part of the savings, but the net cost to the Navy remains the same without further decisions.[4]

The Navy must implement other changes, and we suggest that these take place beginning in the fourth year following implementation, after the transition costs have been paid. Training studies[5] have shown that productivity improvement is achievable from having fully trained personnel in units. This improvement stems from having junior personnel who can perform directly the bulk of needed tasks more quickly and from having more senior personnel having to provide less formal and informal on-the-job training and direct supervision. Moreover, in this analysis, the E-5s and E-6s who previously were in school are now in operational billets in lieu of E-2 sand E-3s, who are now those in school. By itself, this should lead to a decrease in required operational billets. These savings have been assessed in a range of 5 to 15 percent of manpower in previous studies. This suggests that a real billet reduction of a minimum of 500 billets is possible. We extend our analysis by taking a very conservative approach and estimat-

---

[4]   Also, we recognize that the savings are in the Military Personnel account and any training costs are in the Operations account.

[5]   These studies are reviewed in Appendix E of Thie et al., 2009. In particular, see studies cited there by Thomas Manacapilli and Stan Horowitz.

ing the impact of reducing the overall need for IT billets by 200. If these billets were removed proportionally across all grades (from the IT community), there is a direct annual cost avoidance of approximately $20 million to the IT community. This could become an actual Navy savings if the billets were used to reduce end strength rather than distributed to other communities.[6] However, another strategy to pursue with the increased productivity is to keep the billets and costs the same but achieve improved effectiveness for services to the fleet and other IT customers.

Sensitivity of Assumptions

As stated earlier, this analysis relies on assumptions. The first assumption is that the one-time cost to revise curriculum for an audience with five years of service to an audience with one year of service is minimal or moderate. In other words, current plans to revise communications C school curriculum will need to occur regardless of the year of service of the attending students. However, even if the entire curriculum cost (currently estimated at $60 million) would be required only to permit earlier attendance at C school, our analysis suggests only minimal effect on the comparison of cases over time, given the magnitude of the steady-state savings.

The second, and perhaps the key, assumption is the retention rates of 6YOS personnel relative to 4YOS personnel.[7] If 6YOS retention rates were considerably worse than 4YOS retention rates, the relative benefit of the cases would differ. As stated previously, the 6YOS retention rates used in the cases presented fell within the range of actual recent retention rates for a relevant 6YOS community (ETs). Nonetheless, we provide here additional findings to explore the effect of changing the retention assumptions considerably. Alternative B changes the retention at six years and at ten years from 50 percent and 47 percent, respectively, to 40 percent and 40 percent. Alternative C changes the retention even further, to 35 and 40 percent at these key decision points.

Figure D.5 repeats Figure D.1 with Alternative B and C retention assumptions.

Figure D.6 is a repeat of Figure D.4, with all retention cases included. It shows that with either of the lower retention assumptions, at worst, cost is roughly comparable and the benefit of a fully trained IT workforce remains for the Case 3 excursions.

---

[6]   If the Navy is unwilling to gain the productivity benefit, the analysis simplifies to an investment cost up front to gain the additional trained IT. The cost is approximately $25 million per year for three years for Case 3.

[7]   Note that if both 6YOS and 4YOS retention changed, such as to reflect a civilian economic shift, then the differences would be minimal.

Figure D.5
Analysis with Alternative Retention Assumptions

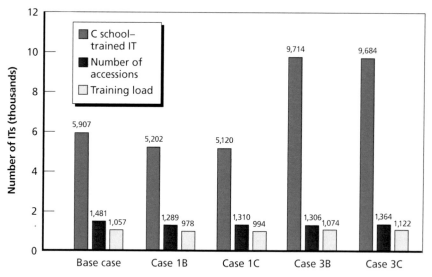

RAND *MG896-D.5*

Figure D.6
Analysis with All Retention Cases

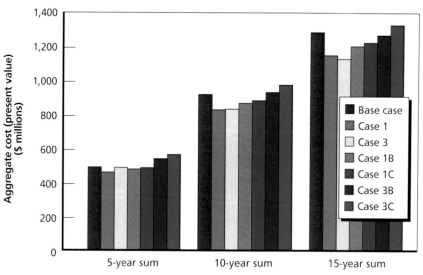

RAND *MG896-D.6*

Costs and Benefits of Earlier C School When Considering Effectiveness Reductions

The analysis suggests that earlier C school provides significant benefit when the metric is the number of trained personnel. The costs of these changes are twofold. First, there are considerable transition costs that will be apparent while the system is sending both early career and later career personnel to C school. Second, there may be steady-state costs compared with the base case. While our initial analysis suggests that there may, more likely, be steady-state savings, this analysis is dependent on retention assumptions. Should the retention difference between 4YOS and 6YOS personnel be greater than initially estimated, the relative cost of early C school will also be greater than initially estimated.

However, earlier training of 6YOS personnel can permit a reduction in overall IT personnel. If we extend our analysis by incorporating even a modest 200-billet reduction, discussed previously, as a feedback loop into our steady-state analysis and using the lowest retention for 6YOS personnel, the savings are apparent. This analysis is reflected in Figure D.7. The base case is the same as previously shown. Case 3C-200ES shows the cost implications of reducing end strength by 200 people, given the 6YOS program that places only trained ITs in units and thus increases effectiveness. This case was calculated with the lowest retention assumptions, but still indicates the significant savings of accessing, training, and compensating 200 fewer personnel.

Figure D.7
Analysis with Assumed Billet Reduction

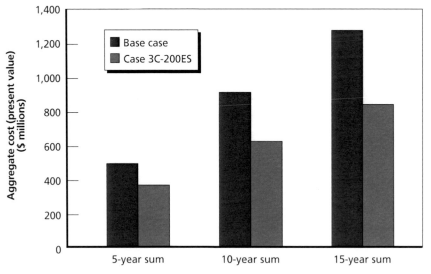

Further Discussion

This is a rough order of magnitude analysis using our steady-state assessment of the training options along with reasonable assumptions about transition effects and cost. We recognize that many stakeholders have cognizance over this analysis and may challenge assumptions and analysis. We present it as a useful way to have a discussion among the stakeholders.

References

Bresnahan, Timothy F., Erik Brynjolfsson, and Lorin M. Hitt, "Information Technology, Workplace Organization, and the Demand for Skilled Labor: Firm-Level Evidence," *Quarterly Journal of Economics*, Vol. 117, No. 1, February 2002, pp. 339–376.

Chan, Stephen L., "Information Technology in Business Processes," *Business Process Management Journal*, Vol. 6, No. 3, 2000, pp. 224–237.

Chang, Yongsung, and Jay H. Hong, "On the Employment Effect of Technology: Evidence from US Manufacturing for 1958–1996," *Penn Institute for Economic Research Working Paper 03-004*, 2003. As of September 24, 2009:
http://pier.econ.upenn.edu/WorkingPapers/2003.htm

"Computer-Based Failure," *Navy Times*, June 15, 2009, pp. 24–26.

Department of Defense, Directive 8570.1, "Information Assurance Training, Certification, and Workforce Management," August 15, 2004.

———, Instruction 8500.2, "Information Assurance (IA) Implementation," February 6, 2003. As of May 8, 2008:
http://www.dtic.mil/whs/directives/corres/pdf/850002p.pdf

Department of the Navy, Memorandum: Navy Tactical Command Support System (NTCSS), Navy Training System Plan (E-70-9099), April 8, 2002.

———, Memorandum: Integrated Shipboard Network System (ISNS), Navy Training System Plan N6-NTSP-E-70-0304A, June 8, 2005.

Dertouzos, James N., and Steven Garber, *Human Resource Management and Army Recruiting: Analyses of Policy Options*, Santa Monica, Calif.: RAND Corporation, MG-433-A, 2006. As of September 29, 2009:
http://www.rand.org/pubs/monographs/MG433/

Gali, Jordi, "Technology, Employment, and the Business Cycle: Do Technology Shocks Explain Aggregate Fluctuations?" *American Economic Review,* Vol. 89, No. 1, March 1999, pp. 249–271.

Garcia, Federico E., James L. Gasch, and Mitzi L. Wertheim, *Workforce Assessment of Information Technology Sailors*, CNA (CRM D0006070.A2), July 2002.

Hoewing, Gerald L., "Manning the 21st Century from a Position of Strength," *The Hook*, Fall 2004. As of September 29, 2009:
http://www.tailhook.org/FA04_Hoewing.htm

Klein, Gary, Laura Militello, and Beth Crandall, "Case Studies Related to Manning Reduction," in Patricia Hamburger, ed., *Integrated Command Environments* (proceedings volume), *Proceedings of SPIE*, Vol. 4126, 2000, pp. 209–220.

Koopman, Martha E., and Heidi L. W. Golding, *Optimal Manning and Technological Change*, Center for Naval Analyses Research Memorandum 99-59, July 1999.

McGovern, Wayne, *Ship/Fleet Manpower Document Development Procedures Manual*, Navy Manpower Analysis Center, January 2005.

Moore, Carol S., Anita U. Hattiangadi, Sicilia G. Thomas, and James Gasch, *Inside the Black Box: Assessing the Navy's Manpower Requirements Process*, Center for Naval Analyses Alexandria, Defense Technical Information Center, CNA report CRM D0005206.A2, 2002.

Naval Sea Systems Command, Program Executive Office Ships (PEOS), *Manpower Estimate Report for the Littoral Combat Ship (LCS)*, October 2007.

Pinsonneault, Alain, and Kenneth L. Kraemer, "Middle Management Downsizing: An Empirical Investigation of the Impact of Information Technology, *Management Science,* May 1997.

———, "Exploring the Role of Information Technology in Organizational Downsizing: A Tale of Two American Cities," *Organization Science,* Vol. 13, No. 2, March–April 2002, pp. 191–208.

Sims, William H., "Potential for Ship Manning Reductions from New Information Systems Technology," CNA Annotated Briefing 96-114, January 1997.

Stoloff, Peter H., William H. Sims, David L. Reese, and James L. Gasch, *NEC Utilization Study*, CNA (CAB D0014616.A4), September 2006.

Stymne, Bengt, Jan Löwstedt, and C. Patrick Fleenor, "A Model for Relating Technology, Organization and Employment Level: A Study of the Impact of Computerization in the Swedish Insurance Industry," *New Technology, Work and Employment*, Vol. 1, No. 2, September 1986, pp. 113–126.

Thie, Harry J., Sheila Nataraj Kirby, Adam C. Resnick, Thomas Manacapilli, Daniel Gershwin, Andrew Baxter, and Roland J. Yardley, *Enhancing Interoperability Among Enlisted Medical Personnel*, Santa Monica, Calif.: RAND Corporation, MG-774-OSD, 2009. As of September 29, 2009: http://www.rand.org/pubs/monographs/MG774/